Arthur Henry Heath

A manual on lime and cement, their treatment and use in construction

Arthur Henry Heath

A manual on lime and cement, their treatment and use in construction

ISBN/EAN: 9783337214715

Printed in Europe, USA, Canada, Australia, Japan

Cover: Foto ©berggeist007 / pixelio.de

More available books at **www.hansebooks.com**

LIME AND CEMENT

LONDON:
PRINTED BY WILLIAM CLOWES AND SONS, LIMITED,
STAMFORD STREET AND CHARING CROSS.

A MANUAL

ON

LIME AND CEMENT

*THEIR TREATMENT AND USE
IN CONSTRUCTION*

BY

A. H. HEATH

London:
E. & F. N. SPON, 125 STRAND
New York:
SPON & CHAMBERLAIN, 12 CORTLANDT STREET
1893

PREFACE.

A REVIEW of a book on Portland Cement ended with these words: "The latest information on a dry and dusty subject." The adjectives may seem somewhat disparaging of the subject, yet the writer was evidently a man of far wiser discernment than to intend giving encouragement to aspiring souls who consider such information as insignificant detail, not worthy of serious attention. Those engaged in constructive work can hardly fail to recognise the importance of a thorough knowledge of cementing materials: it needs no proof, and may be briefly summed in the paraphrase, "No good cement, no good building."

The present book is an endeavour to present to its readers a useful manual, derived from the principal authorities, on the subject of limes and cements, and it is hoped that they may not find in it much that will have to be un-learned. The pages are purposely not cumbered with footnotes or with more references than seem to be needed. The improvement of the book will not be neglected.

The author is greatly indebted to a colleague, Dr. F. E. MATTHEWS, F.I.C., for his kind assistance in revising the chemical portions, for important suggestions, and for his careful supervision of the proof sheets; to Prof. McLEOD, F.R.S., and to others, for valuable information and suggestions.

A. H. HEATH.

THE ROYAL INDIAN ENGINEERING COLLEGE,
    COOPER'S HILL.

CONTENTS.

CHAPTER I.

PAGE

Definition of lime and cement—Rocks furnishing lime—Slaking of lime—Use of mortar—Properties of pure lime mortar and of hydraulic lime mortar—Substances conferring the property of hydraulicity on pure limes—Pozzolana—Trass—Santorin earth—Volcanic ash of the Vivarais district—Aden—Catania —Influence on pure lime of soluble silica—Tests for hydraulicity—Sale of lime and sand 1

CHAPTER II.

Limestones used for production of lime—The upper, the gray, and the lower chalk beds—Chalk marl—The oölitic limestones —The lias limestones—The magnesian limestones—The carboniferous limestones—Gypsum—Selenitic lime—The calcination of limestone—Blast furnace slag cement; its manufacture and use—Chaux de Teil—Ciment Grapier—Chalk and lias limes in France—Roman cement—Keene's and Martin's cement—Parian cement—Lime, &c., in India—Use of surkhi —Volcanic ash from Aden—Molasses—Sugar syrup 16

CHAPTER III.

Portland cement—Definition—Analysis of average sample—Locality of cement works—Analysis of chalks—of clays—Proportions of chalk and clay—Analysis of clay, slurry, and cement—Process of manufacture of cement—Reduction of chalk and clay to liquid mud—Drying and calcining the mud—Temperature of calcination, and its products—Ransome's kiln—Stokes' continuous calcination process—The Dietsch kiln—The phenomena of burning Portland cement—Influence of acid lining of kiln—Influence of temperature—Unloading the kiln and selection of the clinker—Grinding of clinker—Storage of cement powder—Packing of cement for delivery—French burr millstones—High temperature of ground powder—Edge runner and roller grinding mills—Appearance of calcined clinker—Periodical testing of liquid mud—Methods of testing—Proportion of ingredients of Portland cement—Magnesia in cement—Detection of magnesia in cement—Influence of soluble silica in Portland cement—Of alumina—Proportions of silica and of alumina in Portland cement—Chemical changes in the setting of Portland cement—Notes on treatment on some set cements—Uses of Portland cement   42

CHAPTER IV.

Tests for Portland cement—Aëration—Storage—Selection of samples for testing—Mixing of cement mortar—Moulding of briquettes for testing—Shape of briquettes—Immersion of briquettes—Testing of briquettes—Test loads—Unwin's formula for ascertaining approximate tensile resistance—Standard sand—Periodical slurry tests—Statement of average composition—Objection to test of mortar briquettes—Hot-water bath—Deval system of hot-water immersion—Test loads for briquettes subjected to hot-water bath—Test for fineness of grinding—Sieves used—French and German sieves—Thickness of wire—Specific gravity or density test—German tests—A French specification—American practice in cement testing

—Simple tests without use of tensile stress machine—Adhesion test—Mann's adhesion test and machine—Stevenson's transverse bending test—Test of pats of cement in hot water—Contraction test—Expansion test—Prussian test for constancy of volume—Adulteration of cement—Acceptance and storage of cement—Rejection and removal of cement 79

CHAPTER V.

Consideration of tests of Portland cement—Weight and fineness of grinding—Examination of Portland cement mortar under a microscope—Testing by tensile stress—Ratio between tensile and compressive resistance—Objection to use of sand-mortar test briquettes—Density test—Diminution of density with age—Schumann's method—Chemical analysis : its value and capability—Influence of silica and alumina—Quick-setting Portland cement—Needle test for setting—Vicat's needle test —German needle test—The re-burning of damaged cement.. 105

CHAPTER VI.

Lime and cement mortar—Ingredients of mortar—The mixing of mortar—Proportion of water—Of sand—Reasons for the use of sand—Quality of sand—Various natural sands—Artificial sands—Effect of lime on grains of sands—Granulated slag —Crushed cinder—Proportion of sand used—Quantity of water —Grouting—Larrying—Blue lias lime mortar—Mortar-mixing by hand—By machine—Edge runner mills—Grinding the sand —The lime—Any siliceous addition—French mixing cylinder —Strength of lime mortar—Use of cement mortar—Use of loamy and of fine sand—Addition of sugar syrup—Proportions of sand and water—Mortar in sea-water—Impermeable mortar —Mortar-mixing in time of frost—Plastering—Cement facing and pointing of joints—Cement and lime mixed mortar—Strength of cement mortar—Adhesion of cement mortar .. 116

CHAPTER VII.

Concrete—Stones, sand, and cement—Volume of interstices—Dense and open textured concrete—Proportion of sand to be separately specified—quality and sizes of stones used—River and pit ballast—Broken bricks, &c.—Burnt clay, &c.—Preparation of granulated slag—Mixing of concrete—Hand mixing—Measuring ingredients—Mixing gang—Proportion of ingredients—Quantity of water used—Thin walls of concrete for houses, &c.—Poor concrete for pockets—Concrete for a bridge—For sewers, culverts, &c.—Watertight coating to poor concrete—Effect of age on Portland cement concrete—Machine mixing of concrete—Stoney's—Carey and Latham's—Lee and Co.'s—Messent's mixers—Other mixers—Coignet's mixer—Béton Coignet—Blue lias lime concrete in sea water—Liability to expand—General proportions 139

CHAPTER VIII.

Cement concrete in sea-water—An instance of deterioration—Proportions of concrete—Quality of cement used—Correct proportion of concrete—Analysis of sea-water—Action of water on lime compounds—Action of water and of lime compounds on concrete—Action of sulphates, magnesium salts, &c.—Action of sea-water percolating through concrete—Theoretical composition of indurated Portland cement—Precautions in using Portland cement concrete in sea-water—Concrete tank for ammoniacal liquor—Watertight coating to concrete — Plastic concrete—Use of sea-water — Remedial measures for decaying concrete—Deposition of concrete—Resistance of concrete to transverse bending—Tipping of concrete—Thickness of layers—Packing and ramming of layers—Uniformity of construction—Incorporation of layers—temporary gaps in concrete—Scum on surface of concrete—Contraction and expansion of concrete—Concrete paving—Concrete walling, plastering and facing—Open joints in concrete retaining walls and in paving—Concrete walling for houses, for arches and for floors—Strength of concrete—Deposition of concrete in water through a tube, in a skip, in monolithic work or in moulded blocks, and in bags—Concrete in framework—Concrete blocks and block setting cranes—Concrete hollow cylinders, their use and method of sinking—Weight of concrete—Effective weight in water—Cavities in deposited concrete exposed to waves—Useful memoranda 166

APPENDIX.

I. Manufacture of kankar lime—Nature of kankar deposit—Kilns for burning—Fuel used—Description of kilns, air flues, &c.—Charging the kiln—Proportions of fuel and kankar—Duration of burning—Grinding the calcined lime—Deterioration of ground lime—Admixture of surkhi—Quality of surkhi—Use of kankar lime under water—Kankar lime concrete 189

II. Test to ascertain amount of lime in a limestone—Apparatus required—Method of testing—Formula to ascertain weight of carbonic acid—Table of pressure of aqueous vapour .. 199

INDEX 205

DIAGRAM SHOWING CONSTRUCTION OF KANKAR LIME KILNS.

LIME AND CEMENT.

CHAPTER I.

DEFINITION; PURE LIMES; POZZOLANA, ETC.

1. LIME, used as a cementing material for constructive purposes, is obtained from natural limestone, to which in some cases is added a siliceous, or silico-aluminous, substance.

The term "lime" is generally used to denote limestone material possessing cementing properties to a moderate degree only, while "cement" is reserved for the most energetic and strongest cementing substances; and in the limestone materials the difference is mainly one of degree and not of kind.

2. The rocks utilised for the production of cementing limes are mainly composed of calcium carbonate (carbonate of lime, $CaCO_3$), or of carbonate of lime and silicate of alumina; in some instances a proportion of magnesia is present in combination. The use of these magnesian limestones is generally confined to the locality where they are abundantly found, and where better limes cannot be easily procured.

3. The lime cementing material is prepared by subjecting the natural limestone, in lumps of convenient size, to the action of heat until it is calcined, carbon dioxide (CO_2) being driven off, as well as all combined moisture. The resultant calcium oxide (lime) is reduced to a fine powder, and is mixed with water to form a stiff paste, which is placed between the surfaces to be cemented together. Calcination causes a loss of weight which may amount to nearly one-half that of the raw stone. The loss of carbon dioxide (carbonic acid) from a pure limestone ($CaCO_3$) is 44 per cent.; and that from a magnesian limestone may be still larger. Calcined lime for building purposes can be purchased either in the lump or in a powdered state, and is called quick or caustic lime. Pure calcium oxide has the peculiar property of slaking; when lumps of it are sprinkled with water, combination takes place readily, the lime swells to nearly double its former bulk (in the ratio of 27 to 47), and the chemical action is so intense that a considerable amount of heat is developed, and steam will be generated. The lumps of lime then fall almost instantaneously into a fine powder called slaked lime (CaO, H_2O). A cubic foot of ordinary building lime in lumps will make about $1\frac{1}{6}$ cubic foot of slaked lime powder. This fine powder is quite dry, if the proper proportion only of water has been added, about $\frac{1}{3}$ by weight, or as 18 to 56. Slaking is a rapid chemical combination of calcium oxide and water, forming a hydrated oxide, or calcium hydrate, from which the water can be separated only at a red heat.

4. This property of slaking is most marked with the

purer calcium oxide. When silicate of aluminium (clay) is present in small quantity, the action is less intense and slower; it decreases with the increase of the silicate, until a proportion is reached at which the slaking may, for all practical purposes, be said to cease. Such aluminous limestones, after calcination, must be reduced to a powder by artificial means, such as grinding between millstones, or by crushing.

5. The powdered lime is used as an adhesive and cohesive body by mixing it with just as much water as will yield a stiff paste; this paste is inserted between the surfaces of the two bodies to be cemented together, completely filling the space, and is then left undisturbed to harden into a substance resembling a limestone. When a good cementing lime has set, and hardened, the coherence between the particles of lime, and their adherence to the surfaces of the two bodies, should be so complete that, in breaking the united mass, the line of fracture should pass indifferently through cementing substance and cemented body.

Pure Lime.

6. The chemical change occurring during the process of hardening of pure lime is mainly due to the absorption of carbon dioxide from the atmosphere. When this absorption takes place in the presence of moisture, the conversion of the hydrate into a carbonate is accompanied by a crystallisation, the crystals adhere closely to one another and to any rough surface with which they are in contact. This

is the setting, and induration, of the pure lime mortar.

In pure limes the process of conversion is slow, especially when the area of surface of the lime paste, exposed to the direct action of the air, is small compared with the bulk; and, when the surface is converted, the carbonate forms a coating which hinders the ingress of air to the interior of the mass. Hence in a large mass of materials cemented together with pure lime, the hardening of the central portion may be delayed for a long period, if not entirely stopped. The thickness of the film of carbonate formed on an exposed surface may be from $\frac{1}{4}$ to $\frac{1}{8}$ of an inch in the first year, and the increase lessens in each succeeding year.

Moreover, the crystalline form of pure calcium carbonate is soluble in water containing carbonic acid, and consequently will be gradually washed away if exposed to rain. Hence these pure limes are fit for use only in the form of thin coatings to rough surfaces of solid structures, having a maximum of surface exposed to the air, and at the same time under protection from the weather.

The purest natural limestones contain some insoluble silicates of iron and alumina, varying in quantity from 1 to 6 per cent. Less pure limestones may contain from 15 to 30 per cent. of similar impurities, and are materially affected thereby; they slake more slowly, and with less intense chemical action, the increase in bulk and the heat generated are less than are found with pure limes.

7. Some natural limestones contain a sensible pro-

portion of clayey matter (silicate of alumina). When such a limestone is calcined, reduced to powder either by slaking or by grinding, and used as a cement, it is found to be independent of direct contact with the air for the development of the property of hardening, and, moreover, the indurated body is insoluble in water. Clay is a hydrated aluminium silicate, and when calcined together with calcium carbonate, there is formed calcium silicate and calcium aluminate. This compound body, ground to a fine powder, and moistened with water, becomes hydrated; and, slowly re-arranging its particles, becomes a compact indurated insoluble mass, quite independently of atmospheric agency. These limes therefore are well suited for use as cementing bodies for all constructive purposes; and when the aluminium silicate attains a certain percentage, the cement paste will harden readily even when submerged in still water, and will be perfectly insoluble. Such limes are termed hydraulic limes, and are of great value to the engineer.

If the aluminous limestone does not contain, or is not mixed with, sufficient clayey matter for the complete conversion, after calcination, of the caustic lime into these silicates, a certain amount of the lime will be left uncombined, which when exposed to moisture will become a soluble hydrate and be liable to be washed away. The oxide will also expand in the process of hydration, a result which may be harmful.

8. Limes containing about 8 to 12 per cent. of suitable clayey matter are termed moderately hydraulic. When water is added to the calcined lumps, they

break up to a small extent, and heat but little, the slaking is incomplete. When used as a cement the paste or mortar sets under still water after the lapse of 15 to 20 days, but never becomes hard. It hardens satisfactorily when used for building above ground.

Limes containing 15 to 18 per cent. of clay may be termed hydraulic. Slaking begins about an hour after the wetting of the calcined lumps, and is incomplete; the increase in bulk is small. A cement paste sets after 6 to 8 days' immersion in still water, and finally hardens to the consistency of a soft stone.

Eminently hydraulic limes contain from 20 to 30 per cent. of clayey matter. The action of slaking in these limes is delayed, and the lumps of calcined lime do not readily crumble into powder, sometimes not at all. A mortar paste hardens on the third or fourth day of immersion in still water, and continues to harden for a long period; after the lapse of a month it is hard enough to resist the action of running water.

9. The property of hydraulicity may be conferred on pure limes by the admixture of substances containing soluble silicates. A volcanic ash, called pozzolana, (consisting of about 50 per cent. of silicic acid, 16 of alumina, 12 of oxide of iron, 9 of lime, and small portions of other substances), added to pure lime, confers hydraulicity, and the resultant mortar may be used in engineering construction. Smeaton used Aberthaw infra-lias lime, and pozzolana shipped from Civita Vecchia, in the construction of the Eddystone lighthouse in 1756–9.

Pozzolana is found near Pozzuoli, on the west of

Naples, at other places on the flanks of Mount Vesuvius, and probably at other ancient volcanic craters in the district between Naples and Rome; it is a volcanic ash. It is partly powder, partly coarse grained, or like pumice stone scoriæ or tufa stone, and the colour ranges from white, whitish gray, blackish gray, brown, to violet red. The best is said to be the white to blackish gray.

It is largely used in the district, and in Rome; sometimes it is mixed with a pure lime only, sometimes the fine powder is mixed with sand in equal bulk. About 30 parts of lime are mixed with 70 parts (by volume) of pozzolana, consisting of both powder and small stones, which serve instead of sand; the mortar is used for brickwork. The mixture of equal bulk of sand and pozzolana powder can be used for hydraulic work.

Pozzolana differs much in quality, analyses give the following range of composition :—

	From	Average
Silica	44 to 56	47·7
Alumina	10 ,, 15	14·3
Magnesia	0 ,, 4	3·9
Sesquioxide of iron	7 ,, 29	10·3
Lime	1 ,, 10	7·7
Alkalies, &c.	5 ,, 15	4·2
Sand	0 ,, 5	2·5
Water, &c.	0 ,, 9	7·0

The silica in pozzolana readily gelatinises when treated with weak acid, and is in a condition to combine with caustic lime on its hydration, forming a durable silicate.

⊬ 10. A similar volcanic ash is found in the Eifel district, between Bonn and Andernach, on the west bank of the Rhine, and is called "trass"; and another volcanic ash used in the Mediterranean is known as Santorin earth, from the cresent-shaped island of Santorin, in the Greek Archipelago, situate about 75 miles north of Crete. These substances are probably molten lava blown out of a crater, in the form of dust and small scoriæ, by violent gaseous explosions. Trass is largely used in Holland in fresh-water and marine submerged engineering work; and Santorin earth has a great reputation for conferring hydraulicity on rich limes.

Analyses of trass from Andernach give proportions as follows:—

Silica	46 to	57
Alumina	20 „	14
Magnesia	1 „	7
Iron oxide	5 „	5
Lime	2 „	11
Potash } Soda	8 „	15·5

Similar volcanic ash is stated to be procured in central France (the Vivarais district); also at Aden (Red Sea); and doubtless at many other centres of ancient and modern volcanic activity.

A summary of an investigation into the properties of the volcanic ash of the Eifel district is to be found in Proc. Inst. C.E., vol. xcix. p. 407. The results are highly favourable to the use of such ash, containing a large proportion of soluble silica, with either Portland cement, hydraulic or rich limes.

10a. It is stated that in the neighbourhood of Catania, on the east coast of Sicily, and south of Mount Etna, there are beds of clay which have been covered by deep streams of molten lava, with the result that the clay has been burnt and converted into a small red gravel, or powder. This material, mixed with a little quicklime and water, furnishes a mortar which has been used for centuries with excellent results. The buildings in Catania are stated to be constructed of stones of lava cemented together with this mortar, which is also used as an external, and internal, plaster. The lava stones are stated to be small and irregular in shape, but so excellent is the mortar that house walls are built four and six stories high of this material, which is, in fact, a lava rubble concrete. It has also been used in the construction of a harbour breakwater extending for $\frac{3}{4}$ of a mile out to sea, the small lava rubble being moulded into large blocks liberally cemented together with the mortar. It is possible that the clay beds were originally volcanic dust ejected from Etna, or are derived from the decomposition of igneous rocks.

Santorin Earth.

11. The island of Santorin is composed of basaltic and trachytic rocks, at many places covered to a depth of 50 feet with ashes and lava. The ashes constituting Santorin earth occur in the form of a coarse, sharp-grained powder of a light ash-gray colour, with small fragments of pumice stone, obsidian, and other volcanic products.

An average analysis of the earth is stated to be—

Lime	2 to 3	per cent.
Silica	65 „ 69	„
Alumina	13 „ 16	„
Oxide of iron	3 „ 5	„
Magnesia	1 „ 2	„
Potash, soda, &c.	7 „ 8	„

A comparison shows that there is more silica in the Santorin earth than in trass, or pozzolana. It is mixed with lime and sand to form a mortar.

For the Mole works at Fiume the proportions used were—

		Cubic contents after mixing and setting.
6 cubic feet Santorin earth		3·42
2 „ slaked lime		} 2·58
1 „ sea sand		
9 cubic feet		6·00

The Santorin earth, in setting, contracts to the extent of 0·57 of its bulk, hence the difference in volumes of the ingredients before, and after, the mixing and setting. This mixture was employed in moulding blocks which were not to be submerged till 24 to 36 hours after being made. If the mortar is to be used under water when newly mixed, more lime should be added. The above mixture would form a strongly cohesive mass in from six to eight days. For walls of moderate height and width, say 6 to 9 feet, the frames may be removed in from two to four weeks. In heavier works, with walls 24 feet thick, the frames are left in position for five or six months, as the setting of the interior of such large masses proceeds slowly.

Sharp-edged clean broken stone is the best for béton, and the Santorin earth should be washed free from earthy particles.

	Cubic feet.	
Broken stone	19	
Santorin earth	17·50	used at Pola,
Slaked lime	6·31	per cubic yard.
Sand	1·65	

12. Finely powdered burnt brick, provided that soluble silica be present (that is, the bricks have been made of a plastic greasy clay and not of a sandy clay), will also confer this property of hydraulicity on pure limes. The burnt brick, or burnt clay, must be thoroughly and uniformly burned, at least to the point of incipient vitrifaction; and the crushing must be carried to the finest possible state of division; the finer the particles, the more effective is the result of the admixture. The caustic lime must also be reduced by crushing, or by thorough slaking, to a similar degree of fineness. The subject is referred to in the Appendix, on the burning of lime and making of mortar in India. A method sometimes adopted consists in mixing together chalk and a plastic greasy clay in certain proportions; the ingredients are dried and crushed to powder, and then mixed together. Sometimes they are reduced to a moist paste by grinding together in a mill, then the mixture is calcined; this, however, is an imperfect method of making Portland cement, which is fully dealt with later on.

13. On the Continent these expedients are still adopted to a limited extent, in Great Britain they are

rarely used, owing to the comparative cheapness of, and the excellent results obtained from, Portland cement. In distant countries, where neither good natural limes, nor Portland cement, are to be easily procured, these mixtures of the purer limes and the various natural silicates may be used with advantage. The addition of soluble silicates to a cement, or lime, appears to be always beneficial.

Silica, when heated to bright redness in the presence of potash, soda or some alkaline salts, acquires this soluble form. The addition of alkaline soluble silicate and water probably induces a progressive reaction between the silica and the caustic lime, forming a hydrated colloidal calcium silicate (gelatinous silicate of lime), which will solidify and indurate. An interesting hypothesis is suggested by Dr. Matthews, that the alkaline constituents of a cement, or of a mixture of pozzolana, &c., and caustic lime, may play an important part in facilitating the transference of silica to the lime, and the subsequent formation of the gelatinous silicate of lime. The proportion of alkalies in pozzolana, Santorin earth, &c., is noteworthy.

The chemical action induced by the mere mechanical admixture of these silicates is probably imperfect when the process of calcining the mixture is omitted. There is always a certain amount of risk attending the use of a mechanical mixture, due to the possibility of imperfect subsequent combination; the mixing must be perfect, and the ingredients must be in an extremely fine state of division to secure good results. In many instances it may prove to be a good method to mix powdered quicklime with the silicate of alumina, or siliceous volcanic ash, and to calcine the compound. The lime then undergoes a second calcination in contact with the silicate, and the chemical combination during hydration takes place under favourable conditions.

Tests for Hydraulicity.

14. Berthier's test for hydraulicity is stated as follows: the stone is to be powdered, and then sifted through a fine silk sieve to reject coarse particles. Ten grammes of the fine dust are put into a capsule, and by degrees dilute hydrochloric acid is poured on while stirring continually with a glass or wooden rod till effervescence ceases. Evaporate the solution, by gentle heat, to a paste; then mix with half a litre of water and filter, the clay will remain in the filter. Desiccate as perfectly as possible, and weigh. For magnesia, add lime water to some of the above solution so long as any precipitation takes place. Collect precipitate as quickly as possible on a filter, then desiccate and weigh. (See Arts. 19, 20, 55.)

A practical test is to calcine 2 or 3 cubic inches of the limestone in a crucible, crush the calcined lumps to a fine powder, make into a stiff paste with water, and mould in the hands into a ball, which is at once immersed in water. If the lime be hydraulic, the ball will harden to resist the pressure of the finger in about 24 hours, or less; if eminently hydraulic, in an hour or two.

SALE OF LIME AND SAND.

15. In the London district chalk lime is sold by weight and by measure. About 2 cubic yards of ordinary chalk lime in lump make a ton weight; and 16 heaped bushels in the lump make a cubic yard, and 14 heaped bushels of ground lime powder; this proportion is variable.

A measure of lump lime, a London yard, or load, of lump lime is 27 cubic feet, and contains a very little more than 21 striked bushels. A cubic foot is about 0·78 of a striked imperial bushel, and this bushel is 1·283 cubic feet. A hundred of lime is 100 pecks, or 25 bushels. The bushel measures used for lime and cements have sometimes a local value, and also a trade value; the latter is often a given weight of the particular cement; it is best to refer only to the imperial striked bushel.

The London yard of lump lime is from 21 to 22 striked bushels, and makes 18 striked bushels of ground lime (nine 2-bushel bags). A small surplus, for waste, is often allowed in the measure. The local Winchester striked bushel is 2105·42, or 2256, cubic inches capacity.

The approximate weights of quicklimes are:—

Gray chalk lime (ground) .. 40 to 47 lbs. per cub. foot
Blue lias lime (,,) .. 49 to 70 lbs. ,, ,, ,,

The values vary with the quality, condition, and age, of the lime.

Lias lime is also sold by weight and by measure, in

the lump, and in the ground state, in 2-bushel, and in 3-bushel bags. Of ground lias lime, one ton makes ten 3-bushel bags (at the rate of $58\frac{1}{4}$ lbs. weight per cubic foot); of Portland cement, of ordinary quality, about 20 bushels make a ton weight.

Sand is also sold by weight and by measure, about $18\frac{1}{2}$ striked bushels make a ton of dry sand, but the measure is variable, as there are many different qualities of sand.

Another proportion is—about $23\frac{1}{2}$ to 25 cubic feet of dry sand make a ton weight. The cubic yard, and the striked imperial bushels are convenient measures; 21·05 bushels equal the cubic yard nearly, and dry sand is taken as weighing about $88\frac{1}{2}$ to 95 lbs. per cubic foot.

CHAPTER II.

NATURAL LIMESTONES; CHALK; LIAS LIMESTONES, ETC.; GYPSUM; CALCINATION OF LIMESTONE; SLAG CEMENT.

Limestones used in Lime-making. The Upper, Middle, and Lower Chalk, Etc.

16. THE upper or white chalk is found plentifully over a large area in the south-east, south, and south-west of England, and in the eastern counties; in a belt or tract of country running from Dorsetshire to Cambridgeshire and Norfolk, thence to the north of Lincolnshire, and into Yorkshire as far north as Flamborough Head. A small local exposure is at Harefield to the north of Uxbridge.

It is carbonate of lime, containing from 1 to 6 per cent. of silicates and other bodies. A pure calcium carbonate contains 56 per cent. of lime to nearly 44 per cent. of carbonic acid.

When calcined, it loses nearly half its weight, and becomes caustic lime, fit for use, when mixed with water to a paste, in plastering thin coats on rough surfaces protected from the weather. It is called a rich or fat lime.

The upper chalk contains from 55 to 56 per cent. lime, 41 to 43 of carbonic acid, and small quantities of clay, sand, magnesia, and iron oxides.

The gray chalk of the lower chalk division is also carbonate of lime, containing a little alumina (from 5 to 15 per cent. of silica, iron oxide and alumina); when the silicate of alumina is above 8 per cent., the lime is good for cementing purposes. It is found in the same localities as, and beneath, the white chalk; and it is known in the London district and the southern countries as "Gray stone Medway," "Halling" (Rochester), "Wouldham," "Burham"(Kent); "Merstham," "Dorking" or "Guildford" lime. Sometimes it is specified simply as "stone" lime; the term probably arises from a fallacy of Marcus Vitruvius (date about 50 B.C.), that the harder the stone, the stronger the cement made from the calcined stone. Smeaton states that, possibly, these gray chalk limes were styled "stone" limes to convey the impression they were from hard limestone rock. It is also called "flare" lime, from the manner of burning. Gray chalk limes in slaking do not greatly increase in bulk, and only a moderate heat is generated; the time of slaking is longer, and the chemical action is less intense than with white chalk limes.

The lumps of calcined lime are heaped up in a hollow formed in sand, the heap is wetted, and is then covered with dry sand, and sometimes also with sacks, to keep in the heat and assist the slaking. This lime is a good cementing material for ordinary constructive purposes above ground.

The chalk is generally burnt in "flare" kilns, which may hold about 56 cubic yards of lime, and may consume about 8 to 9 tons of coal. In flare kilns, the fuel is not mixed with the stone, but is placed on grate bars, inserted in the floor of the kiln, over which chalk, or stone, is built in an open jointed arch to form a flue, and the flames from the fuel can rise through the joints of the arch. White chalk withstands heat better than the gray; a clayey limestone requires less fuel for burning than the white chalk, but is liable to be overburnt.

The lower chalk marl, of the lower chalk division, is calcium carbonate containing rather more silicate of alumina than the gray chalk, but the proportions are variable and uncertain. When the clayey matter is in excess, the resultant lime mortar shrinks in setting. From its want of uniformity it is not an important source of supply of cement. It is sometimes called "clunch" lime. It has been used near Cambridge, and near Hitchin, in the preparation of Portland cement. Chalk marl near Farnham is stated to contain as much as 21 per cent. of silica and alumina.

17. In the Oölitic series are many beds of limestone; some, interspersed between the marls of the Kimmeridge clay, also those in the Oxford clay, may be found to yield a good hydraulic lime, but it is advisable to test the stone, both by analysis, and by experiment, before using for any important purpose. In Dorset, the Kimmeridge clay cement stones, found east of Weymouth, may perhaps have given rise to the name of Portland cement; the same stones are found near Wareham. A good lime is furnished by some much fissured limestone beds near Peterborough, also near Stamford, the lime is called "blue and gray stone lime." In the Corallian beds of the Lower Oölite are the clayey limestone beds in the Vale of

Pickering, and along the Howardian Hills, in Yorkshire; they are called cement stones, or "throstler."

In Yorkshire also, but of the Lower Greensand formation, are the septaria nodules of the Speeton clay, north of Flamborough Head; these stones were sent to Hull for cement-making.

Where the proportion of alumina in the clayey matter is excessive, the mortar made of the lime is liable to crumble away on exposure to the weather; if the silica be insufficient to convert the caustic lime into a silicate, the mortar is liable to expand in setting, and to produce cracks in the brickwork. To counteract this tendency, care must be taken to ensure perfect slaking of the quicklime before use.

The great beds of Oölitic limestone supplying the valuable building stones are generally not of themselves capable of furnishing a good quality of cementing material.

			P.c. silica.	P.c. alumina and iron oxide.
Portland stone	contains about		1·20 ..	—
Ancaster	,,	,,	1 to 2 ..	—
Ham Hill (Somerset)	,,	,,	4·70 ,,	8·30
Barnack	,,	,,	0 to 2 ..	1·30
Bath	,,	,,	1 to 2 ..	—
Chilmark or Tisbury.	,,	,,	10·4 ..	2·00

LIAS LIMES.

18. In the Lias series are thin beds of limestone separated by seams of clay. These stones contain silicate of alumina in the proportion of from 10 to 30 per cent., with carbonate of lime, and small proportions of other ingredients.

Analyses give as an average composition of selected stone—

Carbonate of lime	68 to 80 per cent.
Silica	20 to 10 ,,
Magnesia	about $1\frac{1}{2}$,,
Alumina	about 3 to 4 ,,

An excellent cementing material is procured from these lias beds at Lyme Regis (Dorset), Watchet, Keynsham, Street, and Pylle (Somerset); also at Westbury, and at Charlton, near Shepton Mallet; at many places in Gloucestershire; from the infra-lias beds of the Aberthaw Promontory (Glamorgan), and at Lliswerry, near Newport (Monmouth). In Warwickshire, the district around Harbury (near Leamington, on the G.W.R.), also Kineton, Wilmcote, and other places near Stratford-on-Avon; at Stockton near Rugby; at Barrow-on-Soar, and elsewhere, in Leicestershire; and northwards to Long Bennington, about midway between Grantham and Newark; and near Kirton Lindsey in Lincolnshire, are some of the places where the lime is produced. The localities mentioned indicate that all along the outcrop of the lias beds, extending from the south coast of Dorsetshire to Lincolnshire, this useful limestone is to be found plentifully, though the quality may not be of uniform excellency for cementing purposes. In North Lincolnshire and in Yorkshire the character of the beds appears to alter.

There are also small patches of lias beds near Whitchurch, in Shropshire; near Carlisle; and in Antrim and in Argyllshire.

It must always be borne in mind that permanence

in the proportions of the ingredients of a natural limestone is not to be expected; not only do the different seams of stone vary widely in the character of their composition, but also there is no certainty of permanent composition in any seam.

The quicklime produced by calcination of selected lias limestone is eminently hydraulic, its faculty of hardening when submerged in water grows with the increase in the proportion of silica. It gives good results as a cement for all ordinary work, either above ground, or beneath in damp situations, or even in still water.

As an instance of the use of lias lime in an important work, Smeaton adopted the infra-lias lime from Aberthaw on the south coast of Glamorgan, about 12 miles south-west of Cardiff, mixed with pozzolana shipped from Civita Vecchia, in the construction (in 1756–9) of his lighthouse on the tidal submerged Eddystone rock, about 9 miles from the coast.

The beds of limestone are thin, and commonly much fissured, and the natural surface of the stone is brownish yellow in colour, owing to the action of moisture. The interior of a block is generally bright blue on a fresh fracture (due to iron protoxide), hence it is commonly styled "blue lias limestone."

The stone is calcined in open topped cup-shaped kilns, the fuel is interstratified with the limestone, about 6 to 7 tons of coal are burnt to produce 25 tons of lime.

The calcined stone must be thoroughly ground to a fine powder, then spread in a layer of about 12 inches thick on the floor of a dry weatherproof shed, and exposed to the air for two or three weeks before it is used in building. A good plan is to fit up the aëration shed with shelves, or trays, on which the lime can be placed.

Lias lime can be slaked in the ordinary manner, but ample time must be given—not less than three days, and the slaked lime must be sifted through a fine-meshed sieve, to reject all coarse, imperfectly slaked particles.

If used when freshly ground, it is liable to expand when hardening in the mortar joints, and cause rupture in the masonry. The expansion is due to the delayed slaking of impure caustic lime.

Aberthaw lime is still largely used in engineering works, a recent instance was in the construction of the Avonmouth Docks, both for lime concrete and for mortar for the masonry.

An analysis of Aberthaw limestone gives—

Calcium carbonate	about	86·0 per cent.
Magnesium ,,	,,	2·0 ,,
Silica	,,	8·0 ,,
Alumina	,,	1·0 ,,
Iron oxide	,,	2·0 ,,
Water	,,	1·0 ,,

An excellent "Portland" cement is now made from the lias limestones, and the clay of the beds separating the seams of stone. If due care be taken to mix the lime and clay in correct proportions, according to their respective analyses, the cement will be good; but the variableness of composition of each ingredient renders the manufacture more difficult than with pure chalk and clay.

Magnesian Limestone.

19. The term magnesian limestone is applied to a carbonate of lime containing upwards of 10 per cent. of magnesia; and is also given to a well-defined geological formation beginning in the south near Nottingham, and extending northwards, in a comparatively narrow outcrop, to the sea at Tynemouth.

Some limestones of this geological formation contain, however, only a small proportion of magnesia; for instance, the Marsden, and Fulwell (Durham), magnesian limestone contains about $1\frac{1}{4}$ per cent. of magnesia to about 98 per cent. of calcium carbonate.

The stone occurs plentifully in the Permian formation in the Midland and Yorkshire districts in England, and is used in blocks for building purposes. In some instances it yields a fairly good cementing material, more or less hydraulic according to the percentage of aluminium silicate present in the stone.

At Portishead there is a magnesian limestone containing about 53 per cent. of calcium carbonate and about 37 of magnesium carbonate, and from near Bristol, on the N.W., a similar stone containing about $1\frac{1}{2}$ per cent. of silica. The Bulwell lime is largely used in the Nottingham district.

Magnesian limestone must be burned at a high temperature to develop its hydraulic properties; if the temperature be low the caustic lime absorbs water slowly, and seems devoid of hydraulic properties.

A pure magnesian cement hardens in water, but a

limestone containing a large percentage of magnesia is likely to be untrustworthy, as the caustic lime absorbs water rapidly, while caustic magnesia absorbs slowly, so that the lime may have set in a mortar and be hardening while the magnesia is undergoing hydration, and its subsequent setting may cause disintegration of the mortar.

Magnesia may be looked upon as an adulterant of the limestone, and must be considered as requiring a similar proportion of soluble silica to confer hydraulic properties. Limestones containing from 42 to 52 per cent. of calcium carbonate, and from 42 to $53\frac{1}{2}$ of magnesium carbonate, with from 4 to $5\frac{1}{2}$ and upwards of silicate of alumina, yielded a hydraulic lime.

20. The Rosendale cement, so largely used for engineering works in the United States, is derived from an argillaceous magnesian limestone of the Appalachian range. It is stated that there are 17 distinct layers of limestone, which vary considerably in quality. The stone is burned, mixed with coal, in continuous draw kilns, part of the charge being withdrawn every 12 hours. The calcined lumps are carefully picked over, those underburnt being returned to the kiln. The lumps are first cracked in a roller-mill and then ground between stones, the fineness being about 95 per cent. to pass through a 2500 mesh sieve. It is packed in paper-lined casks.

The average tensile resistance of the neat cement is stated to be, at 7 days, 104 lbs. per square inch, and at 30 days, 134 lbs.; and of 1 cement to 1 sand, 102 lbs. per square inch in 30 days.

Carboniferous Limestone.

21. The limestone formation known by the geological term of carboniferous limestone is composed mainly of calcium carbonate. There is usually in the stone a small percentage of aluminium silicate, from about 2 to 10 per cent., and in some cases, where the percentage is comparatively large, a good cementing material is produced.

It is burnt in high open-topped kilns, about 1 ton of coal is burned to produce 4 tons of lime. These kilns are sometimes called continuous or "draw" kilns, the firegrate being at some height above the base of the kiln. As the burnt lime descends to the base below the firegrate, it can be withdrawn, while fresh stone is added at the top. The kiln contains from 80 to 100 tons of limestone.

The carboniferous limestone is very widely distributed, forming large parts of the ranges of hills in the north and west of England, and in Wales. One of the best localities for this lime as a cementing material is at the Halkin Mountains, near Holywell, in Flintshire, whence supplies were drawn for the building of dock walls, &c., at Liverpool. The stone is stated to contain about 72 per cent. of calcium carbonate, about 20 of silica, and $3\frac{1}{2}$ of alumina.

In general the carboniferous limestone is suitable for making mortar to be used above ground, and in dry situations; its use is mainly confined to the locality in which it is found. Many of the beds of stone are highly fossiliferous, the fossils being largely composed of phosphate of lime, consequently the calcined lime is uncertain in quality, and has a tendency to slake unevenly, unless well mixed in the powdered state. The fossils may be considered as an adulteration in a more or less pure calcium carbonate.

22. In the Silurian and Devonian geological systems are numerous beds of limestone; some are used for fluxing purposes in the smelting of iron ores, others for constructive purposes, where accessibility and favourable conditions of working and of transport permit.

Among these beds are the Wenlock limestones, quarried near Dudley and Walsall, the Woolhope limestone; the Ludlow, Aymestry, Bala, and Llandeilo limestones, in the Silurian system; and the Petherwin limestone in the north-east of Cornwall, also at Barnstaple, Ilfracombe, and Lynton, of the Devonian system.

	Walsall Limestone, Upper Silurian.	Woolhope Limestone, Upper Silurian.
	per cent.	per cent.
Carbonate of calcium	$76\frac{1}{2}$	67 to 76
,, ,, magnesium	2	2 to 3
Silica	13	14 to 20
Alumina	—	4 to 6
Iron oxide	$2\frac{1}{2}$	—

Such limestones may be found to yield a fairly good cementing material.

Gypsum.

23. Gypsum is a calcium sulphate in a crystalline form, found in beds and layers of the Triassic and Permian formations in England, in the red marl of the New Red Sandstone, and of the Magnesian Limestone. It is extensively quarried on a line of country ranging from the Isle of Axholme (Linc.), by Newark-on-Trent, Elton, and Orston, by Chellaston (near Derby), to Tutbury near Uttoxeter; also in Cheshire, and between Whitehaven and Carlisle in Cumberland. It is also worked in the Purbeck beds found in the Sub-Wealden exploration boring near Battle (Sussex). Other localities are Aust (Gloucester); Droitwich; Syston (Leicester); Watchet and Somerton (Somerset).

It has a normal composition of about 47 per cent. of sulphuric acid, 33 per cent. of lime, and 20 to 21 per cent. of water, hence it is a hydrated calcium sulphate ($CaSO_4, 2H_2O$).

It is quarried either in open workings, or from galleries driven underground, and is heated in kilns to a low temperature, so that all the water of combination is not driven away. When calcined at 120° to 130° C. (248° to 266° F.) it loses three-fourths of its water; it becomes anhydrous at 160° to 170° C. (320° to 338° F.) The heat of calcination must not be too high, as anhydrous gypsum mixed with water sets badly, or not at all.

The calcined mineral, ground to a fine powder, is known as "plaster of Paris," from the original place of production, where it is plentifully found of good quality, and is largely used.

When made into a creamy paste by mixing with water, it hardens rapidly, the water being re-absorbed, and the hydrated sulphate being again formed with some augmentation of volume. It is largely used for plastering surfaces protected from the weather, as it is not insoluble in water, and also for making ornamental castings of all kinds. The dry powder must be carefully protected from access of damp atmosphere; it absorbs water with great avidity.

It is occasionally used for mixing with cement, to make the latter more easily spread, or trowelled, by a workman; or to cause slower setting; but the presence of any considerable quantity of gypsum affects the permanence of strength of mortars and cements, and it should not be used in important work.

It is not of service as a cement for works under water, or exposed to the weather; it adheres firmly at first and sets rapidly, but its adhesive strength diminishes with age. It may be used in a structure which is to be coated with Portland cement, or for the coating of internal walls protected from the weather. The formation of sulphate of calcium in a cement is dangerous to the stability of the compound.

Smeaton is stated to have used plaster of Paris as a temporary cover to exposed edges of lias lime and pozzolana mortar joints; it cannot withstand the continued action of water, especially of sea water.

Selenitic Lime.

24. Selenitic lime, or cement, is the name given to an artificial mixture of gray chalk, or other similar lime, and a proportion of plaster of Paris.

In one method of preparation, lightly calcined gray chalk lime is re-heated to bright redness in shallow kilns having perforated floors, under which are placed pots full of sulphur. The heat igniting the sulphur, fumes of sulphurous acid rise and form a coating, probably of sulphite, on the lumps of lime, which are subsequently ground to powder, and sifted to exclude coarse particles; the exposure to air converts the sulphite into a sulphate.

In another method, sulphuric acid is sprinkled on the calcined lime, or plaster of Paris is mixed with the ground lime.

This selenitic lime is ground to a creamy paste, water being added; and then the sand may be added to make a mortar.

One proportion is, 4 lbs. plaster of Paris, mixed in half a pail of water, to be added to 1 bushel of lime in the mortar-mixing mill, with sufficient water to make a creamy paste.

It has been used for mortar in the proportion of 6 bushels of sand to the foregoing quantity of selenitic lime paste; and for concrete work, 9 bushels of ordinary gravel.

The best selenitic lime is made from a blue lias lime with the usual admixture of plaster of Paris; and mortar may be made of 3 or 4 of sand to 1 of cement, the lime being first mixed to a creamy paste. It is said that an addition of one-fourth part of good Portland cement gives improved results, which is not at all improbable.

The use this selenitic mortar or cement is not general, it sets with rapidity and soon becomes hard, but is not suitable for use in sea water, nor to be exposed to the weather.

Calcination of Limestone.

25. Limestone is sometimes calcined in the open air in heaps built up of alternate layers of stone and fuel, the sides and part of the top of the heap being thickly coated with clay, to prevent loss of heat and to secure some degree of uniformity in burning. This method wastes so much fuel that it is seldom employed, also the ashes from the fuel are liable to be mixed up with the lime. The kilns commonly used are funnel-shaped with open top; sometimes cylindrical and open-topped; or cylindrical, with a truncated conical chimney; or like two truncated cones placed base to base. The first and second are generally built of the limestone itself, the structure being thickly plastered over with clay inside and out. The third and fourth may be built, for permanence, of firebrick set in fireclay for the inner lining of 14 inches, and of a fire-resisting brick or stone for the outside.

In one method of calcining in kilns the stone and fuel are deposited in alternate layers, and the kilns are styled "running" kilns, as the addition of stone and fuel may follow or "run" with regularity until the kiln is allowed to cool down to be emptied for repairs. Another method is to construct at the base of the kiln a fire-hearth or hearths, arched over with the larger blocks of stone; the fuel is thus mainly kept separate from the stone, which is calcined by the flames and heated gases; these kilns are styled "flare" kilns. The latter are to be preferred for uniformity of calcination.

The dimensions recommended for kilns are, height twice the largest diameter, for flare kilns; for running kilns, three times, four, and even five times. Diameter of orifice of chimney of flare kiln, one-third the largest diameter, hearth opening about one-fourth. Diameter of top of funnel-shaped kiln, five times the diameter of the lower orifice, which is usually about 20 inches. Thick walls retain heat and economise fuel. The fuel should contain as little sulphur as possible; if the fuel be sulphurous, calcium sulphate may be formed, and as it sets more rapidly than the caustic lime, and is decomposed both in the open air and in sea water, it may cause disintegration of any masonry cemented with such lime.

BLAST FURNACE SLAG CEMENT.

26. In the iron-making districts, the slag or cinder from the blast furnaces, an aluminous calcium silicate, is sometimes used in the production of a cement.

The slag must be carefully selected to contain the proper proportion of silica, alumina, and lime. As it runs in a molten state from the furnace, it is led into water, or on to a mechanical spreader working above water, or is allowed to flow in a small stream downwards in front of a nozzle from which issues a jet of air or steam. It is thus reduced to light frothy vesicular lumps, sand or dust, or to a fibrous state; the latter, called slag wool, is preferred at the Skinningrove Ironworks, in the Cleveland district. The slag is then crushed or ground to a fine powder, and is carefully screened.

Then a quantity of pure slaked lime, in the form of a fine dry powder (from chalk or any pure limestone, calcined, crushed, slaked to powder, and sifted) is added to the ground slag, and the two are intimately mixed by grinding to a very fine state of division, and are again screened. The sieve test for final grinding is a residue not exceeding 25 per cent. on a 32,000 mesh sieve; usually the residue is 15 per cent. The proportions of slag powder and lime vary according to the composition of the slag, and are measured by weight. Calcination of the mixture before the final grinding is not adopted, and is not expected to be advantageous.

The slag used at Skinningrove is a double silicate of lime and alumina, containing about 30 to 32 per cent. of silica, 30 to 33 of lime, and about 25 to 28 per cent. of alumina. The proportions of lime and slag vary according to the composition of the slag, the total lime in the cement being usually from 45 to 50 per cent.

When slag wool is used, the slaked lime can be added at once, and the two are intimately mixed at one grinding operation. About 25 per cent. by weight of slaked lime is the usual proportion.

BLAST FURNACE SLAG CEMENT.

The resultant cement powder has an average percentage composition, lime 45 to 47, silica 24 to 26, alumina and iron oxide 20 to 22; or lime $4\frac{1}{2}$ to silica $2\frac{1}{2}$ to alumina 2, the lime being about equal to the silica and alumina together. Comparing this with Portland cement, which is generally 59 to 61 of lime, 21 to 23 of silica, 7 to 11 of alumina, &c., the excess of lime and alumina is noteworthy. The colour of the cement powder is a grayish white. The weight of the cement is about 95 lbs. per striked bushel.

For exposure to sea waves a concrete is made of 1 cement by volume to 3 of stone (broken vitreous slag is used) and $1\frac{1}{2}$ of clean sea-sand; for deposition under water, 1 cement to $2\frac{1}{2}$ stone, and $1\frac{1}{2}$ of sand and gravel, the sand alone being equal in bulk to the cement; the maximum proportions for hearting work being 1 cement to 6 of stones and 2 of sand. This proportion is also used for blocks to be hardened in air before use.

For monolithic work, with or without a facing of stronger concrete, the proportions may be 1 cement to 5 stones to $1\frac{1}{2}$ sand; it is better to use the superior facing.

Under tensile stress, the resistance of mortar briquettes made of 3 of prepared sand to 1 of cement is, at 28 days after mixing (27 in water), 370 to 390 lbs. per square inch, on inch section briquettes; increasing in 5 months to about 470 lbs. The limestone generally used at Skinningrove is procured from the magnesian limestone formation of Durham, and comes from a bed beneath the magnesian stone; it contains only about $1\frac{1}{2}$ per cent. of magnesia, and about 96 to 98 per cent. of carbonate of lime. Fulwell, near Sunderland; Tuthill, near Hartlepool; and Raisby Hill, near Ferryhill, are localities of some of the quarries. The tensile resistance of the mortar briquettes is found to follow closely the proportion of residue left on the 32,000 mesh sieve; the smaller the proportion, the stronger is the cement.

This slag cement has been used in the construction of a breakwater and pier about 850 feet long, on an exposed portion of the north Yorkshire coast, at Skinningrove Ironworks. The monolithic mass has up to the present time withstood all attacks of the sea, and shows no signs of any deterioration.

The cement is stated to be essentially hydraulic, and not well adapted for work exposed to air and dryness, as it has been found that the skin of slag concrete so exposed is liable to become disintegrated. It is a slow-setting cement, and shows a tendency to contract during setting. At the Skinningrove pier the concrete below water level has acquired a close, durable, hard surface, which resists attrition and also any decomposing action. Above water, after a lengthy exposure to air, the skin is said to become white and to be less satisfactory than Portland cement in a similar situation; the depth to which the skin is said to be affected is less than 1 inch. This alleged defect was not noticeable on a general inspection of the work. The colour of the dry concrete is whiter than Portland cement concrete, owing to the light gray colour of the slag used for the cement, and for making concrete. The concrete appears at present (1891) to be of excellent quality both above and below water. The manufacture of slag cement is summarised in an article in Proc. Inst. C.E., vol. xcviii. p. 419. It is also dealt with by Mr. G. R. Redgrave, in Proc. Inst. C.E., vol. cv. p. 215.

27. Chaux de Theil, or Teil, is a natural cement or hydraulic lime which has been largely used by French engineers for concrete and other works in the sea, notably for the Suez Canal works (concrete in the harbours of Port Said and Suez). Works built with this cement are stated by Michaelis to have endured for one hundred years in the highly saline waters of the Mediterranean, where some Portland cement works have been destroyed in comparatively few years' time. The composition of this cement is about 62 to 65 per cent. of lime, 15 to $22\frac{1}{2}$ of silica, about $2\frac{1}{2}$ of alumina, $1\frac{1}{2}$ of magnesia, and small quantities of iron oxide, potash, soda and sulphuric acid. It has also been used in the construction of masonry reservoir dam walls near St. Etienne. The proportions used have been about 800 lbs. of lime to about 35 cubic feet of sand, for the harbour works at Trieste.

A somewhat similar cement, styled Ciment Grapier, has a composition of about 61 per cent. of lime to $25\frac{1}{2}$ of silica, 3 of alumina, $1\frac{3}{4}$ of magnesia.

A summary of the qualities and process of manufacture of some chalk limes, and lias limes, in France, is given in Proc. Inst. C.E., vol. xcviii. p. 415.

ROMAN CEMENT.

28. A cement, called Roman cement, was made at the close of the last century from the septaria nodules of the London clay dredged up off the Isle of Sheppey, the Hampshire coast, Folkestone, Harwich, Yarmouth, and also along the Yorkshire coast from Spurn Point to Flamborough Head. These septaria consist of a dark-coloured aluminous limestone traversed by veins—fissures filled with calcareous spar. The nodules were broken up small, calcined in a kiln to incipient vitrifaction, and then crushed or ground to a fine powder.

Similar nodules are found at Boulogne and in the Isle of Wight; and from the Oxford clay, from the lias shales of the Yorkshire coast, and at Weymouth, nodules of aluminous limestone have been treated to yield a similar cement.

The London clay nodules from Sheppey and Harwich contain from 60 to 70 per cent. of calcium carbonate, 18 to 20 of silica, 6 to 10 of alumina, 0 to 2 of magnesia, and a little iron oxide; but their composition is very variable. The stones were calcined in a conical kiln, the clinker crushed and ground in the same manner as Portland cement; the powder must be carefully preserved from contact with damp air to prevent rapid deterioration. Light weight, or low ratio of density is said to be a sign of good Roman cement; the specific gravity of the natural stone is given as $2 \cdot 16$, and of the calcined lump $1 \cdot 58$. The powder must be very finely ground.

In making mortar, about one-third by volume of water is added to the powder, and the paste is thoroughly well beaten up with trowel, or shovel, in the mixing; the more it is mixed the better and harder becomes the indurated cement. A good Roman cement, used neat, sets in air in from 5 to 15 minutes; under water in from $\frac{1}{4}$ to 1 hour. When mixed with sand, the time of setting becomes 1 hour and upwards, according to the proportion of sand. Messrs. Francis & Son make a Roman, or Medina, cement which, if mixed with hot water, sets in about $1\frac{1}{2}$ minutes, and with cold water in about 10 minutes.

This quick-setting cement is valuable for the protection of mortar joints, and of slow-setting Portland cement concrete from the wash of the waves and of tidal or river currents; a coating $\frac{1}{4}$ inch thick of Roman cement is efficient. It is also used for the same purposes in submerged foundation work, and for coating damp walls. The Medina cement is stated to have a tensile resistance of 179 lbs. on the square inch.

The proportions of cement and sand generally used for foundation work are 3 cement to 2 sand; and for plastering, 2 cement to 3 of sand.

Roman cement sets quickly under water, but does not become so strong, nor so hard, as Portland cement. As the powder soon deteriorates if exposed to air, it should be used as soon as practicable after manufacture.

The use of Roman cement is largely superseded for most purposes by Portland cement, which possesses all its good qualities in an enhanced degree, excepting that of setting with extreme rapidity, and Portland can be made equally quick setting, but with a sacrifice of durability.

29. Some so-called cements are really prepared plasters, such as Keene's, Martin's and others, and Parian cement. Keene's cement is a very good hard plaster for coating walls, and is also used for cornice mouldings, skirtings and floors. Calcined gypsum is soaked in a saturated solution of alum, and is then kiln-burnt a second time and ground to powder. Martin's cement is similar to Keene's, only carbonate of potash is used instead of alum. Parian cement is a mixture of gypsum and borax (borate of soda) in powder; the mixture is calcined and ground to powder, and sometimes again mixed with a solution of alum, and recalcined.

LIME, ETC., IN INDIA.

30. In India there appears to be a dearth of rocks furnishing the highest class of cementing materials.

In the plains, lime is obtained from a precipitated concretionary deposit of impure calcium carbonate called "kunkur" or "kankar," found at shallow depths in the alluvial soil. The nodules, of all sizes, are sometimes beaten with stout sticks to knock off adherent dirt, but would be cleaner if carefully washed free from earthy matter, and are then calcined in a kiln. As the composition of kunkur is variable, the resultant quicklime is little to be trusted for mortar-making until carefully tested. An analysis of kunkur gives a composition of calcium carbonate, 72 ; silica,

15 ; alumina and iron oxide, 11 ; but probably most of the silica is in the form of sand.

Sea-shells are also calcined for lime, and another source of supply is the limestone boulders carried down the beds of streams and torrents during the wet season ; the stones are collected from dry gullies, and are broken up and calcined ; the quality of the limestone is variable. Limestone rock is also used. In some localities a calcareous tufa stone is used for producing quicklime ; and, when the depositing waters have been also charged with siliceous matter, a useful cementing material is obtained.

Clay may be added to the kunkur, or limestone, before burning ; it must be a pure, plastic, unctuous clay, free from sand. A good practice is to add finely powdered clay to the crushed calcined pure lime, to mix thoroughly, and again burn the mixture at a red heat. This is stated to give better results than when raw clay and unburned lime are added together before calcination.

Highly-burned brick, or pottery, crushed to a fine powder, is frequently added to quicklime, and used in the place of sand for mortar making. This addition of aluminium silicate, if the silica be capable of assuming the gelatinous state, will confer the property of hydraulicity on the lime according to the proportion of soluble silica. The brick should be crushed as fine as possible. In the plains of India, sand of a quality suitable for mixing mortar is seldom to be found. It is generally too fine-grained, the texture of surface too smooth, and it is mixed with a great deal of loamy matter.

It is stated that a good hard hydraulic mortar was made of kunkur, burnt clay, calcined ironstone and sand. The same ironstone mixed with lime and calcined, gave a good hydraulic lime. For the burning of kankar lime, and the use of surkhi, see Appendix, p. 193.

For lime concrete a proportion is sometimes adopted of lime and soorkee, or surkhi (brick powder) in equal bulk, and together about 45 to 50 per cent. the bulk of the broken stone, of $\frac{1}{2}$ inch to $1\frac{1}{2}$ inches gauge. That is, 1 of broken stone to $\frac{1}{4}$ lime to $\frac{1}{4}$ surkhi; unslaked lime to be not less than one-sixth of the total bulk of the broken stone, &c. A weak lime concrete for backing may be made of 6 of broken stone to 3 of a mortar composed of 2 sand, or of surkhi, to 1 lime, turned over and mixed together at least seven times. Concrete when deposited to be well rammed, compression about one-fourth.

As kunkur and other limestones are generally burnt with wood or charcoal fuel, the calcined lumps should be clean and free from ashes, all underburnt pieces picked out; and should be delivered within seven days from the date of burning. The ground brick, or pottery (soorkee, or surkhi), must be perfectly clean, free from sand, earthy raw clay, or underburned clay; and must be sieved to a gauge of not less than 4624 meshes to the square inch (68 gauge). The bricks made by the natives in India are small and thin, and are of the sedimentary clayey mud found on the surface of the beds of water tanks and ponds in the dry season. The clay is spread out in a thin layer and subdivided into rectangular flat cakes, which are then burnt with dried cow-dung, or wood fuel. Bricks for surkhi are frequently dug out of the ruins of ancient cities, of which Sirhind is an instance.

31. At and near the western sea-ports of India, volcanic ash and pumice stone is imported from Aden, and mixed with the Indian limes in the proportion of ½ or 1 part of pumice to 1 of lime. The pumice is to be ground to pass through a 50 gauge sieve, and is then added to the lime, and again ground and mixed. It must be used within 48 hours of the time of mixing. Finer grinding would be beneficial.

32. It is customary in India to mix with the lime, whether for mortar or for plastering (chunam), a proportion of coarse sugar syrup, or of molasses (goor or jaggree). The effect is to retard the evaporation of moisture, and it is generally considered that the coarsest syrup assists the setting of the mortar, and increases its strength. For mixing mortar, about ½ lb. of sugar syrup is dissolved in 2 gallons of water. Or, from ⅛ to 1 per cent. by weight of the syrup is added to the lime. Or, 1 lb. of molasses to 1 bushel of Portland cement. Refuse molasses is said to be composed of 49 per cent. of cane sugar, 10 per cent. of carbonate of potash, 18 per cent. vegetable matter and mineral bodies, and 22 per cent. of water. About 35 per cent. by weight of the sugar syrup liquid may be used to the unit of lime, for mortar mixing.

As a test for lime in India, if the mortar-paste sets hard after immersion in water for 7 days, it is considered to be hydraulic; and Vicat's needle test is used to ascertain the energy of hardening.

CHAPTER III.

PORTLAND CEMENT; MANUFACTURE; CALCINATION; GRINDING; INFLUENCE OF MAGNESIA; SILICA, ETC.; USES OF PORTLAND CEMENT.

PORTLAND CEMENT.

33. THE most trustworthy, strongest, and most useful cementing material is produced mainly by calcining an artificial mixture of carbonate of lime and clay. It is called Portland cement, not from any connection with the Oölitic limestone quarried in the Isle of Portland, but possibly from the similarity of colour and texture between a surface of fracture of a block of hardened cement and the limestone, and possibly from Smeaton's statement that from Aberthaw lime and pozzolana he did not doubt to make a cement equal to the best merchantable Portland stone in solidity and durability. Or the term may originate from a cement made from nodules found in the Kimmeridge clay near the Isle of Portland.

The average chemical composition of the cement is given by the following range of analyses; also an average sample of a good cement made on the banks of the Thames.

	About	Average.
Lime	58 to 66	$61\frac{1}{2}$
Silica	20 to 26	$22\frac{1}{2}$
Alumina	$2\frac{1}{2}$ to 10	8
Potash	1 to 3	$1\frac{1}{2}$
Soda	0 to 2	

	About	Average
Magnesia	0 to 3	2
Iron oxide	2½ to 4	2½
Carbonic acid	1 to 3	0
Phosphoric acid	1	½
Sulphuric acid	1 to 2	1½
Water	1	0
		100

It is therefore mainly composed of lime with silica, alumina, iron oxide, and some alkalies. A substance yielding this compound is found in the natural state, but the proportions are liable to vary widely. Formerly cement stones (nodules called septaria from the London clay, and Oxford clay) were dredged up off the Hampshire coast, near Chichester harbour, off the Isle of Sheppey, also near Southend, Harwich, and other places, and were calcined to produce the so-called Roman cement. Sheppey cement stones are stated to have a composition of about 64 per cent. of calcium carbonate, 18 of silica, and 7 of alumina, with about 6 of iron oxide, but the proportions are variable.

Of late years all the best cements are made from an artificial mixture of natural substances containing known proportions of the requisite ingredients for cement. As a general statement, a good mixture for cement will contain from 72 to 77 per cent. of calcium carbonate, and from 28 to 23 per cent. of clayey matter. Hence the principal manufactories of Portland cement are to be found where there is an ample and cheap supply of these two bodies, in the form of chalk, or other pure calcium carbonate, and a suitable clay. Sometimes both are to be found in the same locality, chalk in the hills, and the clay on the adjoining river foreshores, or the chalk is brought by barges or ships to the clay deposit, and *vice versâ*.

34. In this country there are cement works in the Isle of Wight; at Northam near Southampton; at Poole and Wareham (Dorset); in Warwickshire and Leicestershire (on the lias formation); on the banks of the River Thames and Medway (on both sides of the Thames from Erith to Gravesend); at Dovercourt (Harwich); at Harefield (Uxbridge, Middlesex); at Arlesey, near Hitchin; near Yarmouth; on the banks of the Humber Estuary near Hull (Stonesferry and Barton-on-Humber); at Hartlepool; on the banks of the River Tyne; and at other places to which chalk can be cheaply conveyed, or where there is abundance of pure limestone and a suitable clay.

On the Continent are large works near Boulogne; at Stettin on the Baltic, where there is chalk and clay in abundance; at Bonn, on the Rhine, and at other places.

Taking the method of manufacture as carried out at works where chalk and clay are used, the following is a description of the process and the machinery in general use on the banks of the Thames.

Either the upper, or the lower, chalk beds are quarried, and the clay is a greasy bluish clay dug out of the alluvial deposits in the ancient beds of the rivers Thames, Medway, and Medina (Isle of Wight).

35. The following are average analyses of the chalk:—

		Upper chalk.	
Calcium carbonate	about	97·90 to	98·60
Silica	,,	0·66 ,,	1·59
Magnesium carbonate	,,	0·10 ,,	0·21
Iron oxide	,,	0·35 ,,	0·74

		Gray chalk.
Calcium carbonate	about	87·35 to 96·52
Silica	,,	1·67 ,, 6·84
Alumina	,,	1·14 ,, 0·93
Magnesium carbonate	,,	0·10 ,, 0·50
Iron oxide	,,	0·38 ,, 0·46
Potash and soda	,,	0·42 ,, 4·29

The composition of the clay varies widely, the following proportions fairly represent a good clay.

Silica	about	55 to 70 per cent.
Alumina	,,	11 ,, 24 ,,
Lime	,,	4 ,, 8 ,,
Magnesia	,,	1 ,, 2 ,,
Iron oxide	,,	3 ,, 15 ,,
Potash and soda	,,	3 ,, 4 ,,
Carbonic acid	,,	4 ,, 5 ,,

Dr. Michaelis gives the composition of a selected clay as, silica, 60 to 62½ per cent.; alumina, 17 to 25; and carbonate of lime, 4 to 9 per cent.; potash and soda, and iron oxide, 2 to 3 per cent. A gault clay contains silica, about 47; alumina, 16; carbonate of lime, 25; iron, alkalies, &c., about 12 per cent.

The analysis of clay must distinguish between the sand (or insoluble), and the soluble (or combined) silica, or it is useless.

The clay should contain very little sand, iron oxide or vegetable matter. The iron oxide renders the mixture of lime and clay more fusible in the heat of the kiln, and is considered to tend to the production of overburnt, vitrified cement clinker, instead of the well-burnt semi-vitrified lumps; moreover it is considered to be a diluent.

The silica of the clay should have a tendency to

gelatinise readily when treated with acids after being burnt. Sandy clays contain a large proportion of insoluble silica which may not undergo a favourable change.

A good cement-mud before burning contains from 68 to 78 per cent. of calcium carbonate, from 21 to 15 of silica, and from 10 to 7 of alumina; consequently the proportions of chalk and clay mixed to make cement, vary according to their respective composition. About 65 to 75 per cent. of chalk, and about 35 to 25 per cent. of clay are, roughly, the quantities used; but careful analyses should be made from time to time of the materials used, and the exact proportions selected to be suitable for the especial class of cement required to be manufactured. A rough measurement sometimes adopted is 3 or $3\frac{1}{2}$ barrowfuls of chalk lumps to 1 barrowful of clay.

The following independent analyses indicate the composition of the clay, the dried " slurry," or mixture of chalk and clay mud, and the cement powder.

	Clay.	Slurry.	Cement.
Lime..	—	—	62·13
Calcium sulphate	—	—	2·13
Calcium carbonate..	2·01	69·97	—
Silica (soluble)	54·14	11·77	20·45
Alumina	14·68	4·45	8·05
Magnesium carbonate	4·48	2·87	—
Magnesia	—	—	1·48
Iron oxide	7·76	2·13	4·37
Sand	0·87	1·24	0·98
Water	15·03	7·59	

36. The chalk and the clay are sometimes dried and ground separately to a fine powder and then are mixed; after being thoroughly mixed dry, sufficient water is added to make a stiff paste, such as can be compressed into cakes, which are then dried and calcined. This method is considered to give a result inferior to that obtained by the reduction of the ingredients to a finely-divided mud.

Formerly both chalk and clay were reduced to a fine mud holding about three times their weight of water; the liquid mud, after passing through a sieve, was run off from the mixing mills to large shallow tanks called settling beds or "backs," where the solid particles settled and consolidated, the clear water above being run off from time to time. The consolidated mud, necessarily deposited in layers of particles ranging from coarse to fine, and of irregular composition of chalk and clay, was dug out after being air-dried for a time, and was then dried on floors, generally heated by the waste heat from the calcining kilns.

The more modern system is to mix only a small quantity of water, about 35 to 40 per cent. of the total weight of the creamy paste or slurry.

Some makers prefer the old system of settling tanks, claiming that better results are obtained by it if fine-meshed sieves are used, and there is less liability to the presence of white specks, indicating the presence of unmixed lime in the calcined clinker. But, on the other hand, a large area of land is required for the settling tanks, and where space for the works is limited, or expensive, the modern process is adopted of pumping the creamy mixture on to artificial drying floors, where the greater part of the moisture is speedily expelled.

37. The reduction of the chalk and clay to an impalpable mud is effected in a disintegrating machine called a wash-mill. It is in the form of a ring, or octagonal trough, about 14 feet diameter, in which are dragged, at about 22 revolutions per minute, framed cutters and tines to break up and mix the materials. In the old fashioned wash-mills the periphery is surrounded with wire gauze having 900 to 1024 meshes to the square inch, permitting only finely-divided mud to escape into a collecting trough.

In some cases measured quantities of clay and chalk are placed in a wash-mill; in others, a separate mill is used for each ingredient; and a measured quantity of liquid mud from each mill is thoroughly mixed in a third mill.

In one modern process the mud as it comes from the wash-mill is ground between two broad horizontal plates of cast iron superposed, and surrounding the wash-mill. The lower plate is fixed; the upper one is capable of rotation, and all the mud escaping from the mill must pass between these grinding plates, and be reduced to a fine state of division. The cast-iron ring plates are about 9 inches wide, and may have plane or slightly serrated faces; they are built up in segments.

In another process, called the Goreham process, the mud escaping from the wash-mill is conducted to a pair of French burr millstones running horizontally, and comes thence as a creamy mud of very fine impalpable nature. The grindstones may be up to 4 feet diameter, and are driven at 120 revolutions per minute.

In some works edge runner mills are employed — large discs of cast iron running on edge in a ring trough ; these are considered to be more economical in driving power than grindstones, and to yield a finer mud than comes from the ordinary wash-mills.

It is important to secure fine grinding, the completest mechanical mixing of the chalk and clay, and to give no opportunity for the injurious separation of these two ingredients of different density.

The finé mud, or slurry, should be sufficiently fine to pass through a 22,500-mesh sieve and leave a residue not exceeding 8 per cent. ; in some modern works the slurry is passed through high-speed centrifugal sieves.

When slurry is imperfectly ground, minute white particles of caustic lime are seen in the burnt clinker.

38. The mixed slurry, as it comes from the washmills, must be frequently tested by a simple analysis to ascertain if the required proportions of lime and clay are present. The usual method is to ascertain the amount of carbonic acid in a measured quantity of slurry, and thus estimate the percentage of lime.

The slurry is then pumped on to the drying floors, generally placed close to the calcining kilns. The floors, of either stone or cast iron, have beneath them flues through which heat from fires, or from the calcining kilns, circulates. In some cases, the heated products of combustion are allowed to pass over the

E

surface of the mud, as well as beneath; in others only surface extraction of the moisture is employed. The latter plan is said to give good results, but there is the danger of the absorption of an injurious amount of sulphur from the fuel.

39. The excess of moisture being driven away, the dried slip (or slurry), containing about 10 per cent. of water, is then broken up into lumps about the size of a man's fist, which are taken to the kiln and burned generally with gas coke fuel. Great care and skill are needed in packing the slip in the kiln, and in conducting the calcination, which must be carried up to, but not beyond, incipient vitrifaction at a white heat. Two forms of kiln are commonly used, the intermittent dome kiln, with open top, or with flues carrying the waste heat to the drying floor; and the continuous ring kiln, built as a number of kiln chambers surrounding a central chimney stack. In the intermittent kiln, the chamber is successively packed, heated, allowed to cool, and unloaded; in the continuous ring kiln all these operations may be in progress at the same time in a succession of chambers. Ring kilns are expensive in construction, but they permit a very complete use of the heat produced in calcining.

An ordinary size of intermittent kiln will be charged with 70 tons of slurry and 15 tons of coke, and will yield about 30 tons of cement clinker and 15 cwt. of ashes. The temperature attained in the kiln should be between 2500° and 3000° F., and the loading, burning, cooling, and unloading takes about 6 or 7 days, distributed as one day each to loading, &c., and

3 to 4 days to burning. To reheat the kiln costs each time about 5 tons of coke.

Caustic lime is formed at about 824° F. (440° C.) and it is considered that at a temperature of about 1290° F. (700° C.) the silica and lime unite. At a higher temperature alumina is supposed to unite with the silicate of calcium. The maximum that should be reached, for a short time only, is about 2912° F. (1600° C.) or say 3000° F. It is stated that the proportion of free caustic lime in the clinker depends upon the deficiency of heat in the kiln below the temperature required to produce semi-fusion.

Thorough and complete burning, rapid calcination at a high temperature, is essential to the production of a good cement; the maximum of hydraulicity in the cement is developed only at an intense heat; that is, the chemical process of combination of lime, silica, and alumina in the setting of cement, is liable to be imperfect unless the required heat has been attained.

Ransome's Kiln.

40. As the burning of the dried cement mud in lumps of irregular size is always more or less imperfect, a novel process of burning has been recently devised by Mr. Ransome, at Grays, Essex. The dried cement mud is reduced to a coarse powder, in which state it is burned in a chamber consisting of a hollow cylinder of wrought-iron or steel plate, lined with fire-brick set in fire-clay. About every fourth course of brick is set on end, making a series of parallel longitudinal projecting feathers, or ridges, extending from end to end

of the interior of the cylinder. On the exterior of the cylinder are three stout hoops, one at each end, and one central; the latter bears a toothed rack gearing into a worm which can be rotated at a slow speed. The end hoops rest on rollers fixed in the supporting framework for the cylinder. The cylinder is mounted on this framework so that its axis is inclined to the horizontal at an angle of about 15° to 20°.

Gas is used as fuel, made in one of the ordinary forms of gas producers for heating purposes (Siemens' or Wilson's). The slurry powder is delivered into the chamber through a hopper placed at the higher end; the gas enters, and is burnt at the lower end, and the flames traverse the length of the chamber. The powder falls into the rotating cylinder, and is continually lifted up by the ridges, and dropped through the flame, and at the same time it moves gradually down to the lower end of the cylinder. A transit occupying about 30 minutes, giving about 50 passages through the flames, is generally sufficient, and the burning goes on continuously, until the cylinder itself requires relining.

The advantages claimed for this method of burning are: that the heat is under complete control, that every particle of the cement is exposed to approximately the same amount of heat, the calcined powder is in a favourable state for grinding, without need for crushing; and the process may be continuous for a long period, till the lining of the cylinder requires repair.

An improvement upon Ransome's process, made by Stokes, has been tried at Arlesey, near Hitchin, otherwise the process has not yet been largely adopted.

Stokes' Continuous Drying and Burning Apparatus for Portland Cement.

41. Thick slurry containing from 30 to 45 per cent. of water is pumped into a longitudinal trough, in which is rotating a hollow cylinder set co-axial with the trough. The cylinder is heated by the passage of waste gases through the interior, and as it rotates on its longitudinal axis, the lower part of the shell, immersed in the slurry, picks up a coating which rapidly becomes dry, and is scraped off the cylinder by a scraper, the outside of the cylinder being ribbed transversely to facilitate both the drying, and the separation of the dried slurry, in the form of small sticks.

The dried slurry falls into a conveyer which delivers it into another rotating cylinder, rotating once in a minute, which is the furnace for burning. It is mounted on a slope, and is lined with fire-brick, the slurry gradually traverses the furnace, being exposed all the time to producer-gas flames, and finally falls into another small rotating cylindrical chamber ($1\frac{1}{2}$ revolutions per minute) set to a contrary slope under the furnace. This is both a heating-chamber for the air supply to the furnace, and a cooling chamber for the burnt clinker, and the clinker finally issues from this chamber in the form of small fragments. The air supply enters at the lower end of the cooling chamber, and reaches the furnace at a temperature of about 900° F.

It is stated that 10 minutes' exposure to a clinkering heat in Stokes's kiln is sufficient to produce a good sound well-burned cement.

42. In Joy's system of burning, a domed kiln is used, and the kiln is charged with a mixture of fuel and wet slurry mud, and the upper surface of the contents of the kiln is covered with wet slurry at any points where flames are to be seen breaking through. As the burning proceeds, and the contents of the kiln settle down, dried slurry is fed into the furnace. The system is stated to give good results, owing to the high temperature maintained with a fair degree of uniformity throughout the kiln, beneath the caked covering of wet slurry. The temperature of carbonic oxide burning in air is said to be about 2000° C. (3632° F.).

The clinker is also said to be more honeycombed, and thus more friable, and more readily crushed than ordinary clinker.

On the Tyne, and near Cambridge, an improved kiln called the Dietsch kiln is used, for which the slurry is dried in small blocks of regular size.

The Dietsch Kiln.

43. At the Beerse Cement Works, near Antwerp, the Dietsch continuous kiln is used in burning the dried slurry. In this kiln the fire-hearth is placed at about mid-height of the kiln; the dried slurry is charged at the top, is heated by the gases and flames escaping from the fire, and after passing the fire-hearth, descends to the base of kiln; on their descent the clinkers are cooled, and the air supply to the fire-

BURNING PORTLAND CEMENT. 55

hearth heated by the passage of the air-currents through the heated clinker; the supply of air being drawn upwards from the base of the kiln. The dried slurry is charged at the top, and the burnt clinker unloaded from the base continuously. At these works the clay used is washed and run off into reservoirs, to separate sand from the clay; and the wet slurry is ground between millstones, and then thoroughly mixed together in a large reservoir. The ground clinker has to pass 75-gauge sieves, with only 5 per cent. residue.

Phenomena of Burning Portland Cement.

44. In burning the dried slurry, the first action is the expulsion of water from the clay, and the expulsion of carbonic acid from the chalk (at about 820° F.) Caustic lime acts upon silicate of alumina (at about 900° to 1300° F.), decomposes it, and forms silicate and aluminate of lime (tricalcium silicate, $3CaO, SiO_2$). Probably exposure for a considerable time to the ordinary kiln temperature is necessary for the complete union of the lime and silica; in a recent improved form of kiln a comparatively short exposure to a relatively high temperature appears to effect the same degree of combination. If the particles of lime and clay in the slurry have been reduced to the finest possible state of division, the time of exposure to heat is greatly reduced from that required for coarsely-washed slurry.

In an ordinary kiln, about $2\frac{1}{2}$ hours at the maximum temperature (about 1600° C. or 2900° F.) is at least required. For slurry in lumps, any heat above that specified appears to render the clinker hard on the outside. It is considered that a properly proportioned slurry is not liable to be overburned in a properly managed gas-fired furnace.

In the *Engineer* of the 10th of June, and 14th and 21st Oct., 1892, it is suggested that acid lining of the kiln, and acid ash of the solid fuel may add materially to the percentage of acid in the cement; and that vitrifaction of the clinker is due to local excess of acid constituents. It is also considered that, at a very high temperature, silica will appropriate the greater part of the lime, and there will be a consequent deficiency of lime for the formation of calcium aluminate, upon which the setting of the cement is considered to depend. It is suggested that a cement-mud burned with gaseous fuel may require a composition different from that generally adopted for calcination with solid fuel. An interesting comparison is given of two differently calcined cements from the same slurry. No. 1 analysis is of a clinker that was completely vitrified; it disintegrated spontaneously, and possessed only feeble setting powers. No. 2 analysis, practically identical, is of well-burnt clinker yielding an excellent cement. (Analyses by Mr. A. H. Hewitt.) The alumina alone was in the proportion of about 9 per cent.

(See article in issue of June 10th, and letters from A. H. Hewitt, W. Stokes, and Messrs. Stanger & Blount; pp. 333 and 356.)

	No. 1.	No. 2.
Lime	59·02	60·40
Silica	21·65	22·55
Alumina and iron oxide	14·75	13·60
Magnesia	2·20	2·45
Sulphuric acid, alkalies, and loss	1·58	1·00
Loss on ignition	0·80	—

UNLOADING THE KILN. SELECTION OF CLINKER.

45. When the lumps of cement mud (dried slurry) have reached the stage of semi-vitrifaction, the intermittent kiln is allowed to cool down, and the burnt clinker is taken out. It is carefully picked over, and all overburnt glassy-looking, and all underburnt yellow, pink, reddish, or dull purple clinkers are rejected, the removal of the underburnt being the more important.

The picked clinkers are then broken up in a stone-breaker machine to the size of a small walnut or a hazel nut (this size is the better), and are then fed into the hopper of the grinding mills, which are generally of French burr millstones running horizontally. From the millstones the cement passes through a fine-meshed sieve (sometimes 1600 meshes to the square inch, but should be much finer), the coarse particles being separated for re-grinding.

The powder should be a light gray in colour, with

a slightly metallic tinge, and with a sharp rough feel, without grit, to the touch.

The fine powder is then spread in thin layers, in a dry weather-proof shed, either on a cement floor, or on trays or shelves fitted up to be slightly tilted when desired.

Each day's make forms a layer; the powder may accumulate to 3 or 4 feet in thickness on the floor, or be placed about 12 inches deep on the trays. After exposure to the air for some days (usually from 7 to 14), for cooling and aëration, the powder is ready for use, and is generally packed for transport in bags, or in casks lined with brown paper or a damp-proof material.

The floor layers of powder are dug through vertically, so as to mix together the product of several days' manufacture; with the same object a number of trays and shelves are simultaneously tilted to discharge their contents on to the floor of the shed.

46. Cement bags are usually of a size to hold 2 bushels, or about 200 to 220 lbs., or $2\frac{1}{2}$ cubic feet nearly, and from ten to eleven of them contain a ton weight of cement powder. For export to France the bags hold 50 kilos, or $110\frac{1}{4}$ lbs. The weight of good cement may range from 75 to 90 lbs. per cubic foot. The barrels in general use hold about 376 lbs.; the best are machine made, and weigh each about 24 lbs. Six of these barrels will hold a ton of cement powder. In France, the net weight of the cement in barrels is about 170 to 190 kilos, or about 375 to 419 lbs. The German cask is 170 kilos, or about 375 lbs. of cement, and the sack 60 kilos, or 132 lbs.

It is considered detrimental to keep cement powder for more than one month after making, unless in a perfectly dry, closed, weather-proof shed.

It has been suggested that in many cases an advantage would be gained by exporting the calcined clinker to be ground to a fine powder at the site of the works, as required. Probably the clinker does not suffer any appreciable injury by exposure to air; it is stated by Mr. Reid that a cargo of clinker was sunk under water for five years, and suffered no appreciable change.

French Burr Millstones.

47. French burr millstones are quarried near Paris, and in the district lying between the rivers Seine and Marne, west of Paris, from a highly siliceous fresh-water limestone, probably deposited from hot springs. The formation is the "calcaire silicieux" of the Upper Eocene, and is probably of the same age as the Bembridge limestone beds of the Isle of Wight. The bluish-white stones are the best, the yellowish in colour the next best, and the reddish are inferior. They should be well seasoned before use.

The millstones are built up in two concentric rings of eight shaped stones in each ring, the joints being finely dressed and close, and are cemented. The blocks are bedded on and are backed by cement concrete, and the millstone is hooped with three or more wrought-iron rings. The blocks round the eye of the millstone should be as large as practicable, and the millstones are built up to a diameter of from $3\frac{1}{2}$ to $4\frac{1}{2}$ feet. The blocks are set on edge, that is, with the planes of deposition vertical; if the stones are too siliceous they do not grind the clinker freely, and if there be too little silica in the stone it wears away rapidly.

The term "burr" originally meant the grooving, or furrowing, cut to a particular pattern, on the millstone surface.

The stones are run at an average velocity at the periphery of 25 feet per second, or $4\frac{1}{2}$ feet diameter millstones are run at 140 revolutions per minute, grinding to a fineness of 8 per cent. residue on a 2500 mesh sieve. The cement comes from the millstones at a temperature of about 150° to 160° F., and about 25 to 32 cwts. of ground cement are delivered by one pair of stones in an hour. It is stated that the power required to drive a pair of stones is about 35 to 40 horses-power for this production, or about 25 to 28 horses-power per ton (Carey, Proc. Inst. C.E., vol. cvii. p. 47).

48. As the high temperature of the cement powder on coming from the millstones clearly indicates a waste of power, edge runner mills are sometimes used for grinding the clinker, though many manufacturers prefer the millstones. The edge runner mills consist of a circular pan, or a ring trough, and resting on the floor of the trough are two or four large cylindrical discs of cast iron on edge, called "runners." The discs are mounted on horizontal axles passing across the central pillar of the machine, or forming horizontal arms on a vertical shaft. A small amount of vertical move-

ment of the discs is permitted. In some cases toothed wheel gearing is applied to the axles, or to the vertical shaft, driving the "runners" to run round in the fixed pan; in other cases the pan itself rotates, and causes the revolution of the stationary runners. The pan is generally driven by a pinion wheel gearing into a toothed rack ring bolted on to the underside of the pan; this method is generally preferred to the direct driving of the runners, as the gearing is better protected, and can be kept more free from dirt and dust.

It is found that a four-runner grinding mill is more economical of power than millstones, the expense being only about 15 horses-power per hour per ton of cement powder, as compared with about 25 to 28 horses-power for the same out-turn from millstones; and the temperature of the powder coming from the edge runners is only 70° F., as compared with 150° to 160° F. from millstones.* Another estimate gives about 107° F. as the difference of temperature of millstone ground, and runner or roller ground, cement powder.

Crushing rollers, made of extremely hard chilled cast iron, are also used; each roller works at a slightly different speed from the others, thus creating a rubbing as well as a crushing action. It is considered that this combined rubbing and crushing action gives the best results.

* Mr. P. Neate, Proc. Inst. C.E., vol. cvii. p. 145.

APPEARANCE OF CALCINED CLINKER.

49. Good clinker resembles lava in texture, is more or less porous, has a greenish-black bronzed hue, is homogeneous in character, and is slightly vitrified.

The clinker must show but slight tendency to form dust in the kiln; if this tendency be marked, the clinker probably contains too much clay, it is of a deep brownish-bronze colour, and as it cools after burning, partially disintegrates into a fine flaky grayish powder, a silicate of lime and alumina, which has no value as a cement. The powder feels soft and smooth to the touch; when used as a cement it hardens speedily, and then disintegrates on continued exposure. Such clinker is comparatively easily ground, so the powder may be made to pass a stringent sieve test without disclosure of its defects.

Highly vitrified clinker has a comparatively unalterable structure, and yields an almost inert cement powder.

Clinker possessing great hardness, having a black glossy lustre, and making little or no dust in the kiln, probably contains an excess of lime. It is hard to grind, gives a bluish-gray powder, harsh to the touch, and when used as a cement has a tendency to "blow," that is, expand in setting.

Some clinker shows on a fractured surface numerous white spots, these are due to coarse particles of lime unmixed with the clay; they increase the quantity of free caustic lime in the cement, and are detrimental.

The presence of iron gives a bluish-black colour; a greenish hue is probably due to a small quantity of manganese.

Underburnt clinker has generally a greenish-gray colour; overburnt clinker may have a dull bluish-black, or reddish, or yellowish colour, if sulphur be present in excess.

Clinker containing an excess of silica is liable to decrepitate on unloading from the kiln; this tendency may be lessened by slow cooling.

Periodical Testing of Slurry from Wash-Mills.

50. It is very essential that the clay and chalk be mixed in proportions adjusted to their own composition and that of the cement to be produced. In many works sufficient attention is not paid to the analyses of the ingredients and their measurement into the wash-mills. At modern works, the slip, or liquid mud, is tested hourly by a simple chemical test occupying about 15 minutes; any departure from the standard is soon detected, remedial measures can be adopted, and a fair uniformity can be ensured.

Methods of Testing Slurry.

51. Scheibler's test is sometimes used to check the proportion of calcium carbonate in the wash-mill slip or slurry; it consists in measuring the quantity of carbonic acid given off when a known weight is

treated with an acid, and from that measurement calculating the quantity of lime and also the amount of clay. The substance must be composed of a single carbonate. Scheibler's method is given in the *Engineer* of Jan. 27, 1882 ; also in the Appendix.

Mr. Reid states that Lunge's new nitrometer process, or Japp's volumetric process, is by some preferred to Dietrich's or Scheibler's calcimeter process. The best apparatus for the Dietrich method is stated to be made by Messrs. Griffin & Son (Faija, Proc. Inst. C.E., vol. cvii. p. 119). The manner of use is described by Mr. Carey on p. 49 of the same vol. cvii.

Proportion of Ingredients of Portland Cement.

52. An essential point in cement-making is that the ingredients be in as finely divided particles as possible, and in suitable proportions. One proportion is stated to be, that the lime must not exceed 3 times the silica and alumina, nor be less than $2\frac{1}{2}$ times the silica. The following are said to be suitable proportions for good cement, lime about 2 to $2\frac{1}{2}$ times the silica and alumina, and lime about $2\cdot8$ to $3\cdot75$ times the silica. It is stated that a cement containing 65 per cent. of lime and 35 per cent. of silica, set hard when used as a mortar, but finally broke up and fell into a white powder. On the other hand, a cement containing $73\cdot7$ per cent. of lime and $29\cdot3$ of silica made a mortar which set hard and did not break up. Mr. Carey gives as good approximate

INGREDIENTS OF PORTLAND CEMENT.

proportions, 6 of lime to 2 of silica to 1 of alumina. But it is not at present possible to define very closely the correct proportions of ingredients, as the chemical combination of the components is more or less perfect according to the degree of heat attained in burning. When the temperature of burning is under absolute control, and when every minute particle of clinker is uniformly and thoroughly burned, the exact proportions of cements may be defined.

53. Cement mud, or slurry, containing too much clay, is liable to fuse in the kiln; and the fused clinker when ground to powder yields an inert substance. Overclayed cement powder is light in weight, brown or yellowish in colour, and when used as a mortar sets quickly, but never thoroughly hardens, and is liable to crumble away on exposure to the weather. Excess of clay increases the rapidity of setting.

Cements rich in alumina set quickly, but may be made slower setting by the addition of about 1 per cent. of gypsum; though due regard must be paid to the sulphur compounds already in the cement, for it is found that sulphuric acid compounds, even in moderate proportion, destroy hydraulic cements.

It is better that the alumina should not exceed the proportion of about 10 per cent. in the cement, otherwise the silica compounds are liable to be overburnt (and rendered inert) at the temperature required to effect combination between the lime and alumina. Some cement-makers state that 7 or 8 per cent. of alumina is a necessary ingredient of good cement. The presence of iron oxide also increases the liability to fusion, with similar injurious results.

54. Slurry containing 65 per cent. and upwards of lime requires excessive heat for calcination; the clinker is hard to grind, and the resultant cement is liable to be treacherous unless all the lime has entered into chemical combination.

Overlimed cement powder is of light gray colour, heavy and dense, and when used as a mortar is liable to "blow," or expand in setting, which may be injurious to masonry, or to concrete. Such cement must not be used in sea water, especially if it contains much free lime; and for other less important works the cement powder must be thoroughly air-slaked before use. An excess of lime retards the setting. Free lime in the cement powder may be due either to excess of lime in the slurry, or to underburning of the clinker, the temperature attained in the kiln being less than that required for the combination of all the lime, silica, and alumina. Most Portland cements contain a little free lime, and it is advisable that the precaution of air-slaking the newly-made powder should never be omitted.

Magnesia in Cement Clinker.

55. In some limestones used in the making of Portland cement there is an appreciable quantity of magnesia, and the important questions arise, what is the influence on the cement of caustic magnesia? and what is considered to be the maximum percentage permissible?

Pure caustic magnesia will combine with water,

MAGNESIA IN CEMENT CLINKER.

and form a hydrate of great hardness, but the setting is accompanied by an increase of volume. Magnesium oxychloride forms a quick-setting cement having a high resistance to tensile stress, but as it is acted on by water, it is practically useless as a cement for constructive purposes. It has been used under the name of Sorel's cement in the manufacture of emery wheels, &c. Magnesium chloride and calcium carbonate are stated to make a strong cement. A strong cement is made of a solution of magnesium chloride (of 80 per cent. strength) 9 parts, magnesia 100 parts, and silicic acid 15 parts; but this cement does not endure when freely exposed to air.

When Portland cement clinker is roasted, lime combines with silica and alumina at a lower temperature than magnesia; the latter may be left uncombined, and in the caustic condition. The hydration of free lime and magnesia proceeds at different rates; thus, in a cement containing free magnesia, the sluggish hydration of the magnesia may cause an expansion, and consequent disintegration of the already indurating mortar or concrete; quick-setting Portland cement has been found to set before the completed hydration of the magnesia.

According to Dr. Michaelis, magnesia in calcined cements may be considered as an adulterant which can be counteracted by an increase in the proportion of lime; and, if this be done, a cement containing about 5 per cent. of magnesia need not be rejected as unfit for use. Dyckerhoff, on the other hand, considers that a cement containing more than 4 per cent. of magnesia is untrustworthy.

Another limit which has been stated is $1\frac{1}{4}$ per cent. of magnesia; undoubtedly the less there be in ordinary cement, the greater is its chance of durability.

The magnesia precipitated in cement concrete submerged in, or exposed to the percolation of, sea water, is dealt with under the heading of "Concrete in Sea Water."

56. Dyckerhoff considers that a cement containing 4 per cent. of magnesia tends to expand more than one with 3 per cent.; but that a retrogression in strength is not of a marked character until a proportion of 5 per cent. is reached. Test pieces containing as high as 21 per cent. magnesia remained sound while kept in air or only immersed at intervals, but when wholly immersed showed strong symptoms of "blowing," or expansion. A good many German cements are stated to be made from magnesian limestones.

In Proc. Inst. C.E., vol. lxxxviii. p. 460, are stated the results of observations on Portland cement containing a large quantity of magnesia (29 per cent.), and $43\frac{1}{2}$ of lime. It is stated that concrete made of such cement was, after a time, sometimes after several years, so affected by the slow hydration of the caustic magnesia that the works were completely destroyed. Another instance of failure of cement containing a high percentage of magnesia is cited in Proc. Inst. C.E., vol. lxxxvii. p. 462.

57. Erdmenger recommends the use of high-pressure steam as a means of detecting an injurious quantity of magnesia in a cement. The usual mortar briquettes are exposed to steam at 15 atmospheres

for ten hours, and at the end of this time they must show no signs of injury, be free from cracks, and not be friable, and their tensile resistance must not fall below 15 kilos per square centimetre (213 lbs. per square inch.) Probably the high temperature of the steam is an important factor, and with the use of steam there is not the liability to dehydration of the mortar which would occur if the briquette were exposed to an equal degree of dry heat.

INFLUENCE OF SOLUBLE SILICA IN PORTLAND CEMENT.

58. The general aspect of the part played by silica in the formation of indurated Portland cement is shown in the manufacture of Ransome's artificial sandstone, of the patent Victoria stone, and of similar products.

In Ransome's original process for making stone, crushed calcined flints were used. Subsequently a silica stone from Farnham, containing about 96 per cent. of silica and a little lime, magnesia, and iron, was used.

The silica stone is ground to a fine powder and mixed, under pressure and heat, with caustic potash or soda solution, forming a viscous fluid; this is thoroughly incorporated with a mixture of quartz-sand and lime, and the compound is moulded into blocks. These are immersed in a calcium chloride bath, with the production of a hard and durable calcium silicate, and soluble sodium chloride which is washed away. Subsequently the calcium chloride bath was omitted. In contact with caustic lime, the silicate of soda is decomposed, silicic acid combines with the lime, and caustic soda is set free. In the patent Victoria stone, a Portland cement containing about 60 per cent. of lime to 23 per cent. of silica, and about 10 to 12 per cent. of alumina, is mixed with water to a paste, incorporated with granite chips, and then moulded into blocks. The blocks are immersed for 8 to 10 days in a bath of silicate of soda solution, with the production of a very hard siliceous cement, uniting together the fragments of stone. There are many similar compounds, of small pebbles or stone chips with cement treated in a silicate bath, and styled by different names, such as granolithic, &c. The concrete is always made as dense as possible.

It is evident that silica plays a most important part in the production of a durable cement; the silica must be in the soluble state, and must readily combine with the lime. Investigation is much needed as to the exact proportions for the best combination of silica and lime for cementing purposes.

59. The influence of alumina is not easily to be distinguished; it combines with a certain portion of the soluble silica which might otherwise be available for the lime ; or, if the heat of the calcination be sufficient, an aluminate of lime is formed which is stated to have hydraulic properties, but is probably not durable. Then comes the question, is not the clay (always containing silica in excess) used for Portland cement making, simply a vehicle for supplying free

silica in a state readily susceptible of combination when calcined with caustic lime? and could not an excellent Portland cement be made of silica and lime only?

Mr. H. Reid mentions two facts bearing upon this. An experiment is mentioned, undertaken by Sir C. Pasley, to make Portland cement out of chalk and slate-dust; and an analysis of the slate-dust presumably used is quoted, showing an excessive proportion of 26 of alumina to 38 per cent. of silica. The experiment failed to produce a durable cement.

In another experiment, flints (containing about 98 per cent. of silica, and very small portions of lime, alumina, and oxide of iron) were calcined and powdered, and mixed with lime; the mixture therefore was practically one of lime and silica only. It is stated that an extremely hard and durable cement was made by adding a fourth part of the powdered calcined flint to one part of slaked lime. Probably a second calcination of the mixed powdered flints and slaked lime would be beneficial.

Dr. Michaelis states that a highly siliceous cement sets slowly but hardens with energy, and becomes very strong.

A very powerful and durable cement, largely used for sea-works in the eastern Mediterranean, is made of caustic lime and Santorin earth containing about 65 per cent. of silica in a soluble state, 15 per cent. of alumina, and 2 per cent. of lime. See also analysis of Chaux de Theil, Art. 27.

60. From these analyses it appears that the alumina should bear but a moderate proportion to the

soluble silica; an excess of alumina is stated to impair the indurating qualities of a cement, although aluminate of lime by itself has good setting properties. The kiln temperature necessary for its formation is higher than that for the production of a silicate of lime; hence if a clay containing an excess of alumina be used for cement-making, the silicate of lime will be overburnt before the whole of the alumina has combined with lime; and if the silicate of lime be rightly burned, the probable result is that there will be in such cement some free lime and free alumina, and it will not be durable.

61. A good Portland cement can be made of any limestone and any siliceous rock, provided the ingredients are in the proportions and in the condition indicated as essential for the formation of the durable silicate of lime. Mr. Reid states that he has made a good cement from the Derbyshire carboniferous limestone and the intrusive igneous rock, locally called toadstone, containing from 43 to 63 per cent. of silica, and 18 to 14 per cent. of alumina, with some calcium and magnesium carbonate, and iron oxides. Some toadstone has an earthy fracture, and decomposes readily on exposure; some portions are more compact and are probably more siliceous. Rowley ragstone, from the Rowley Hills, near Dudley, is similar. Some of the slate deposits may yield a suitable clay; many granites apparently contain the desired proportion of silica and alumina, yet a large part of the silica is in a crystalline insoluble form, and the stone must be roasted and crushed before it is suitable for use in cement-making.

ANALYSIS OF ROWLEY RAG (IGNEOUS ROCK), ETC.

	Rowley Rag Stone.	Kendal and Llangollen Slate.
Silica	46	60 to 61
Alumina	16	18 ,, 20
Iron oxide	$19\frac{1}{2}$	7 ,, 8
Lime	11	1 ,, $1\frac{1}{2}$
Magnesia	$3\frac{1}{4}$	2 ,, 4

CHEMICAL CHANGES IN THE SETTING OF PORTLAND CEMENT.

62. The chemical change caused in the cement-mud by calcination is considered to be (1) the conversion of the calcic carbonate into calcic oxide at a dull red heat; (2) the production, at a temperature approaching fusion, of a chemical combination of silicic acid, with soda and potash, alumina and iron. At a red heat silicic acid has greater affinity for potash and soda, and less affinity for alumina and iron than for lime.

When the calcined cement is reduced to a fine powder, and is mixed into a paste with water, the soluble silicates of potash and soda are decomposed, and the silicic acid combines with the newly-formed hydrate of lime. This silicate of lime partially combines with the other silicates of alumina and iron, forming double hydrated silicates which are practically insoluble.

Another view is expressed as follows: lime rendered caustic by calcination reacts upon the siliceous clay, converting it into a compound easily decomposed by acids. The excess of caustic lime, as well as the compound into which the siliceous clay has been converted, then react upon each other when water is added, and a solid stone-like silicate is formed. The water has a double action ; lime and silicate of alumina, when dry, have little action on each other, but if the mixture be wetted, the water will constantly transfer the lime it has dissolved to the siliceous particles ; then fresh lime will be dissolved and transferred, until the process is completed.

A block of hardened Portland cement-mortar consists largely of calcium silicate. The hardening is the effect of a production of calcium hydrate and hydrated calcium silicate, perhaps in the form of extremely attenuated crystals. (?) The chemical reaction in solidification is materially assisted by extreme fineness of the particles of ground cement. Fine grinding of the cement clinker increases relatively the area of surface of contact of particles ; it assists solution of the lime, and is favourable to the hardening ; the finer the particles, the greater appears to be the length of the crystals of calcium silicate. (See Art. 63).

63. Dr. Michaelis, the eminent German authority on cement, states the hypothesis of the formation of a hydro-silicate of lime during the process of induration of hydraulic limes and cements. Attention is drawn to the important conclusion that in the usage of such limes and cements, if exposure to air and heat causes a loss of the water of hydration the cement will lose its cohesive properties. This loss has been recognised in the case of mortar and concrete made with rich lime and Santorin earth (volcanic ashes and pumice stone). In fact no cement can retain

CHEMICAL CHANGES IN PORTLAND CEMENT. 75

uninjured its cohesion unless it be kept in a slightly damp state.

It is also stated that the behaviour of this hydro-silicate of lime is in every respect like a colloid, that is, hardening after the manner of glue.

It presents the appearance of a swollen gelatinous body, with no apparent tendency to crystallisation ; in no case in which this silicate has been examined has the presence of crystals been conclusively proved. The hydro-silicate behaves like a mineral glue, and, during desiccation, becomes more compact, and up to a certain point, increases in cohesive strength ; if the water of hydration be parted with, disintegration sets in ; the more dense the mortar or concrete, the more capable is it of resisting the loss of water of hydration. A pure hydro-silicate of lime is stated to be destroyed with rapidity ; mortars made of rich lime and volcanic ash, within a short period; hydraulic lime and Roman cement, more slowly ; good dense Portland cement mortar, last of all. Even good Portland cement concrete, if exposed to the weather during a hot dry summer, with no supply of moisture available, will probably be found to be covered with a network of innumerable hair-like cracks, which are the signs of the beginning of disintegration, the natural and inevitable result of the loss of the water of hydration of the cement.

The conclusion to be drawn from the acceptance of Dr. Michaelis' hypothesis is that Portland cement mortar and concrete should be made as dense as possible, at least on the surfaces exposed to dryness, where the thickness of the denser coating to a large mass of more porous material should not be less than 6 to 9 inches. When used as a thin coating, the cement should be worked and compressed with a trowel until a maximum of density is attained, and a smooth compact hard surface produced. The supply of water during the mixing of the concrete should not be stinted, and where it is practicable, cement facing and concrete may with advantage be occasionally saturated with water during very dry weather.

The theory of the setting of cements, notwithstanding all the investigations that have been made, is still somewhat obscure. Probably chemical investigation, conjoined to examination under the microscope of polished thin slices of indurated cement, may tend towards elucidation of the process of setting.

Prof. H. McLeod has kindly furnished the following interesting notes of an investigation of some cements :—

A Portland cement containing—

Soluble silica	19·15 per cent.
Lime	62·8 ,,
Alumina	7·86 ,,
Magnesia	1·04 ,,
Ferric oxide	2·76 ,,
Sulphuric acid	2·06 ,,
Insoluble silica	3·6 ,,

was finely ground and shaken with distilled water for 8 days. 40·5 per cent. of lime was dissolved, 38 per cent. being dissolved in 3 days only. Some of the same cement was mixed with water, allowed to set, and kept in water for 3 days. It was then powdered and shaken with water for 3 days ; about 32 per cent. of lime was dissolved.

A block of hardened cement, of unknown composition, 3 inches by 3 inches by 2 inches, which had been kept in water for $3\frac{1}{4}$ years, was broken ; some of the exterior, and of the interior, was finely powdered and shaken with distilled water for 11 days. From the exterior 26·5 per cent. of lime was dissolved, and from the interior 27·6 per cent. The exterior contained 4·5 per cent. of carbonic acid, and the interior 3·1 per cent.

A briquette which had been tested at 7 days after making, and probably had not been kept in water, was examined $3\frac{1}{3}$ years afterwards. Portions were powdered and shaken with distilled water for 11 days. From the interior portion 36·3 per cent. of lime was dissolved, and from the exterior 23·4 per cent. The interior contained 1·1 per cent. of carbonic acid, and the exterior 8·7 per cent.

Uses of Portland Cement.

64. Portland cement is largely used by engineers and builders in numerous forms of construction, in fact it may be said to be capable of general application in resisting statical pressure, whether vertical, inclined, or horizontal; in the construction of walls of any reasonable height without immoderate increase in thickness over the best masonry ; and in construction of arched bridges of spans up to 26 feet and wider. It is especially suitable for massive and monolithic construction. For foundation work in moulded

blocks of large size for piers and quays ; in monolithic work in dock, quay, harbour and retaining walls ; warehouse and dwelling-house walls ; in abutments of bridges, and in special cases for arches ; in monolithic, or moulded block, hollow cylinders sunk vertically in soft ground for foundations ; in monolithic and moulded block culverts for watercourses and for drainage purposes ; in foot pavements, staircases, and landings ; in weather-proof coating to walls, and in plastering ; in cementing together blocks of stone and bricks, and in lining the interior face of the shell of iron and steel ships ; and sometimes in making steam-tight joints : these are some of the uses of Portland cement. Of late Portland cement sand-concrete (or béton) has been used as a substance in which is embedded either woven steel or wrought iron wire, or woven steel or iron ribbon. The embedding of woven wire, or ribbon, confers great strength on comparatively thin slabs of concrete, and the material can be used in sheets for the covering and protection from fire of structural ironwork in buildings, and for the construction of thin light fire-proof floors, &c. Also for the construction of tanks, and of pipe mains in short length, or in long length without joint. A trelliswork of round bars, with mesh of from 2 to 4 inches, has also been used.

CHAPTER IV.

TEST OF TENSILE RESISTANCE; OF FINENESS OF GRINDING; OF DENSITY; SIMPLE TESTS; ADULTERATION; ACCEPTANCE AND STORAGE.

ENGINEER'S TESTS FOR PORTLAND CEMENT.

65. A BRIEF consideration of the process of manufacture of Portland cement readily leads to the conclusion that when due care is taken in every step of the process, the resultant cement may be admirably suited in every respect for engineering purposes. On the other hand, neglect of precautions and careless management of the process will probably result in the production of an inferior, possibly of a worthless, or even treacherous, material.

66. As a consequence, engineers when purchasing cement from the makers are accustomed to specify certain conditions and tests that are to be satisfied.

The conditions are that the cement is to be newly made, but not too new, not less than 14 days old since being ground, and that the cement-powder has been stored for aëration in layers not more than 12 inches thick on the floor, and on shelves, or on trays, of a weather-proof dry shed. Sometimes the cement is stored 3 or 4 feet deep, in as many separate layers, and when removed for packing, or for delivery in bulk to purchasers, the layers are dug through to the floor.

Aëration should be unnecessary for a perfectly mixed and burned cement, but in the present imperfect condition of manufacturing processes, it is advisable that it be adopted.

The delivery of the cement powder is usually made in 2-bushel bags, or in 400-lb. barrels (net cement 375 lbs.), which should be as air-tight as possible; sometimes they are lined with brown paper as an additional precaution against dampness. A bushel contains 1·283 cubic feet, and may weigh from 100 to 116 lbs. but the weight is variable, according to mode of filling, &c. A barrel of cement contains about $4\frac{1}{6}$ to $4\frac{1}{2}$ cubic feet.

Storage room must be provided for the reception of the cement delivered to the engineer. A perfectly dry floor in a weather-proof shed is essential. The cement-maker may be called upon to provide such a shed, and there store the cement, both for aëration, and during the time occupied in testing its quality; or he may pay a fixed sum per ton for the accommodation provided on the works, but the cement is to be stored in the hired shed at the sole responsibility and risk of the maker, pending the completion of the tests and its consequent acceptance, or rejection, by the engineer.

67. The tests imposed are :—(1) A weight specified for a given bulk of cement, or the density (specific gravity) of the cement is stated. (2) A specified measure of resistance, to gradually applied tensile stress, of a small block (a briquette) of mixed neat cement, and of a cement mortar of specified proportions of ingredients. The shape of the block is

specified and its sectional area at the expected point of rupture. (3) A sieve test of the fineness of the ground powder. (4) A chemical analysis of the cement powder. (5) An adhesion test. (6) An expansion and contraction test.

68. The principal points to be embodied in a specification for the supply of cement for use in engineering work are as follows :—

The cement is to be of uniform quality. At the time of delivery samples will be taken from one sack or package in every five (or in every ten), or from every sack for important work. The cement is to be taken from the bulk of the package, and not from the surface only. In preparing cement for moulding briquettes, Mr. Bevan recommends that the cement powder be placed on slate, or on glass in a ½-inch layer, scored across, and be stored in a cool dry place for 48 hours. About 5 ozs. of cement are required for the usual shape of 1-inch section briquette, and three briquettes should be moulded at the same time.

Test briquettes are to be made under the specified conditions of mixing, moulding, and shape of mould given hereafter. After 24 hours' exposure to air at a temperature of about 50° or 60° F. the test briquettes are to be immersed in pure distilled water for 6, 13, and 27 days, at the same temperature of 50° or 60° F.; and at the expiration of 7, 14, and 28 clear days after the time of moulding, the briquettes are to be broken in an approved tensile stress testing machine, and they are to bear not less than the average load stated for each period in lbs. on the square inch of sectional area.

CONDITIONS OF MOULDING, ETC.

69. The briquettes are to be moulded, in gun-metal moulds of the shape known as Grant's, of a length of 3 inches, of a symmetrical width at each end $1\frac{3}{4}$ inches, of a mid-length width of 1 inch, and uniformly 1 inch in thickness; the mid-length reduction of width being effected by a $\frac{3}{4}$-inch radius of concave curvature on each side, giving 1 square inch of sectional area at mid-length. The radius of the curve joining the concave curvature to the straight side at the end of the briquette may be either $1\frac{1}{2}$, $1\frac{3}{4}$, or $2\frac{1}{2}$ inches. Three briquettes are to be made for testing under each separate condition.

Briquettes of neat cement are to be made for testing at 7 days and 28 days after moulding; and of cement mortar (1 of cement to 3 of standard sand by weight) for testing at 7, 14, and 28 days. Water, in proportion required by the cement, and 50° to 60° F. in temperature, to be added to and mixed with the cement powder. About 25 per cent. by weight of water may be used for the neat cement, and about 15 per cent. by weight, of the sand and cement, for the mortar briquettes. Or, the proper proportion of water for the cement alone is to be taken, and a percentage (about 7 per cent.) added for each part of sand.

The mixing is to be either by hand continuously throughout the testing, or by Faija's mixing machine continuously; machine mixing is the more trustworthy under ordinary circumstances. Faija's mixer, or gauger, is described in Proc. Inst. C.E., vol. lxxv.

Mixing with clean salt water, and mixing in a low temperature, has the effect of retarding the setting of the cement.

The mixed cement is to be lightly placed in the moulds, and then is to be pressed for (5) minutes under a load of (10) lbs. placed on the top of the soft cement projecting above the mould. The loading block to be shaped to the mould, with $\frac{1}{64}$-inch clearance, and is to be placed on the cement symmetrically. After loading, the surplus cement to be cut off with a trowel or knife, and the briquette smoothed level with the surface of the mould. The mould to be filled standing on a horizontal level slab of glass or of glazed earthenware.

Sometimes the moulds, when filled, are vigorously tapped on their sides, to shake down the cement, and make the briquette more dense. Or, the cement mortar is beaten into the mould for one minute with a spatula, weighing half a pound, till the water rises to the top. The loading is suggested as more regular and trustworthy.

In not less than 2, nor more than 6, hours the briquettes are to be released from the moulds, and at 24 hours after moulding are to be wholly immersed in pure distilled water. While in air, the briquettes should be covered with a damp cloth, or be placed in a cloth-lined box, damped to prevent injurious abstraction of moisture.

70. The briquettes are to be fixed for testing in the jaws of the machine in such manner that the line of action of the stress must be identical with the longitudinal axis of the briquette. Double lever machines

for testing are preferable to single lever; the line of tension alters less in direction, and they are more compact.

The loading of the neat cement briquettes may begin at zero, or at 100 lbs., or at half the load specified to be carried; and the rate of increase is to be 100 lbs. per 10 or 15 seconds; for the mortar briquettes the loading is to begin at zero, and to increase at the same rate. A rate of increase of 15 lbs. per square inch of sectional area per second has been recommended in the case of larger briquettes.

The tensile stress which each class of briquette is to bear before breaking is stated in the following table.

For Neat Cement Briquettes.
At 7 days, a stress of 300, 350 or 400 lbs. per square inch.
At 28 „ „ „ „ 480, 550 or 650 „ „ „ „

For Cement Mortar Briquettes.
At 7 days, a stress of 120, 160 or 220 lbs. per square inch.
At 14 „ „ „ about 175, 225 or 290 „ „ „ „
At 28 „ „ „ „ 200, 270 or 350 „ „ „ „

The average value of each group of three briquettes may be taken, provided not more than one of them is more than 10 per cent. below the specified load. Quick loading of the briquettes gives higher results. A three days' test of mortar briquettes (2 in water) should give a tensile resistance of at least 110 lbs. per square inch, for a high class of cement.

A high tensile resistance must accompany a finely ground cement, otherwise the fineness may be due to the grinding of a comparatively soft and under-burned clinker.

Dr. Michaelis considers that a cement which shows

great energy in hardening, that is, attains 90 per cent. of its strength in 7 days, is better than a cement which is more slowly progressive ; on the other hand, a cement giving high results at 7 days has been known to fail in a few months, and some English authorities prefer to adopt a moderate initial strength.

71. The values given in the tables are for ordinary, medium, and high class quality of Portland cement, based partly on past experience, and at the same time looking forward to the general production of a far higher average quality of cement than is now to be procured from some English makers, who have been content to continue using old-fashioned methods and treatment, while some of the German makers have embodied in their practice the most advanced chemical treatment and the highest mechanical skill. The "Stern" brand of cement produced by Messrs. Toepffer, Grawitz & Co., Stettin, has a wide reputation for excellence of quality. The reproach of being behind the age in the making of Portland cement is happily being cast off, owing to the improvements introduced by the more enterprising among English makers. It is probable that owing to recent improvements in manufacture the resistance values may be largely increased in the course of a few years.

Prof. Unwin (in a paper communicated to the Inst. C.E., vol. lxxxiv.) deduces a formula which can be applied to ascertain the approximate tensile resistance of a Portland cement at any time within 2 years after moulding into briquettes, provided that an initial 7 days' value of resistance, and a subsequent value be known. The formula is—

$$y = a + b \sqrt[3]{x - 1}$$

where y is the strength of a cement or mortar at x weeks after mixing; a is the initial strength of that cement or mortar at 7 days, both values in lbs. on the square inch. The constant b can be found from any two values of tensile resistance at known times.

$$b = \frac{y - a}{\sqrt[3]{x - 1}}.$$

Prof. Unwin shows that the calculated resistance at given times does not materially differ from the results of experiments.

Minute details of the best methods of carrying out the testing of Portland cement are given by Mr. Grant, Proc. Inst. C.E., vol. lxii., and in vol. xci. are recent German regulations, specifying the use of a hammer crushing test (Dr. Böhme's hammer test).

French tests for Portland cement are stated in Proc. Inst. C.E., vol. lxxxviii. p. 457.

72. The standard sand to be prepared from crushed quartz rock, and only those grains are to be used which pass through a sieve of 400 meshes to the square inch, and are rejected by a sieve of 900 meshes, the gauge of the wires being No. 30 B.W.G. (0·0122 inch) and No. 31 B.W.G. (0·010 inch) respectively. The crushed sand must be well washed and dried before use.

German standard sand is clean quartz sand passed through a sieve of 387 meshes per square inch, wire to be 0·01496 inch diameter, and then rejected by a sieve of 774 meshes per square inch, made of wire 0·0126 inch diameter. (In German measure the sieves are 60 meshes per square centimetre, wire 0·38 millimetre, and 120 meshes, wire 0·32 millimetre.)

73. Preference will be given to cement works where (hourly) tests are made of the amount of lime in the slurry, and where there is an absolute sieve test for the fineness of reduction of the particles of slurry.

74. A statement is to be furnished by the maker, giving the average composition of the cement proposed to be supplied by him, a variation not exceeding one-twentieth either way to be allowed in any and all of the ingredients of the cement supplied.

75. An objection, sometimes raised against the tensile test of mortar briquettes, is the time ordinarily occupied between the delivery of the cement from the maker, and its acceptance or rejection by the engineer. Also cements have been known to pass satisfactorily a 28-day test, and yet to be bad cements, as shown by subsequent disintegration.

Lately the plan has been adopted of immersing test briquettes and pats in hot water of about 170° or 180° F., or in boiling water, at 24 hours after moulding. It seems that an immersion for 3 days in boiling water may be taken as approximately equivalent to 6 days in cold water, a reduction in the test load of about 14 per cent. for 1 to 3 mortar briquettes being noted in some cases ; but there does not appear to be any fixed ratio, and new test loads for the boiling water testing must be instituted after thorough investigation. Mr. Margetts, of the West Kent Cement Company, states that no less than ten cements of various makes disintegrated more or less under the boiling water test, some being reduced to mud.

The boiling water test seems to be a most powerful agent for the detection of a faulty composition in the slurry, or of imperfect combination of ingredients of a cement; the imperfect combination being largely due to insufficient temperature of burning, and to coarse instead of fine reduction of the ingredients.

Two cements show, on analysis, almost identical composition, one may withstand perfectly the immersion in boiling water, the other may disintegrate; the one has been well mixed and burned, the other is faulty in admixture, and is imperfectly burned.

For neat cement briquettes, boiling for 48 hours, after 24 hours in air, appears to give the briquette a tensile resistance of about 11 per cent. less than after 6 days in cold water.

76. The system of immersing test briquettes in hot or boiling water is sometimes called the Deval system, which may be described as follows:—The cement briquettes, 1 to 3 of sand by weight, are kept in air for 24 hours, and are then immersed in a bath of water heated to 80° C. (176° F.) After 6 days in the hot bath the briquettes are tested by tensile stress, and the results are approximate to those obtained after 27 days' immersion in cold water. Cements which gave doubtful signs of unsoundness at 27 days in cold water, showed symptoms of cracking, blowing, or actual disintegration in the hot bath. In all cases unsound cement was characterised by displaying considerably less strength at 6 days in hot water, than was shown by the same cement after 27 days in cold water. The cement must have completely set before the immersion in hot water—about 24 hours is generally sufficient.

Some experiments mentioned by Mr. Wilfrid Stokes, on 1-inch sand briquettes, give a useful comparison and standard of value.

    7 days' hot test .. 275 to 310 lbs. per square inch.
    28 days' cold test .. 290 to 330 lbs.   „   „

This cement was burnt in Mr. Stokes' new form of kiln, in which thin flakes of dried slurry are exposed to an intense heat for 10 to 15 minutes, practically assuring the complete burning of every particle of cement.

At the same time briquettes were made of cement from the same batch of slurry as above, but burnt and ground to the ordinary commercial standard; these briquettes after the 7 days' hot test bore only 95 lbs. 45 lbs., and 50 lbs. respectively, per square inch.

It is claimed for the hot water bath, that it saves time, that imperfectly combined cements are at once detected, and that a successful resistance proves the soundness of the cement. Mr. Margetts, of Burham, states that the use of the hot bath immersion for 2 days gives almost the same results as 6 days' immersion in cold water.

The cold water bath system often fails in the detection of bad cements, whereas the hot bath always detects the presence of free lime by the lessened strength under tensile stress, and by the cracking, swelling and disintegrating of thin pats. An instance is cited by Mr. Griffith of a cement that passed a 7 days' test (of neat cement), but the briquettes showed signs of disintegration within a month after mixing, the disintegration continuing until there was left only a granular mass.

Test for Fineness of Grinding.

77. The cement clinker is to be ground to such a degree of fineness that it shall all pass through a sieve of 6400 meshes (80 gauge) to the square inch (or a residue of 5 per cent. may be permitted); and also pass through a sieve of 32,400 meshes (180 gauge) to the square inch, leaving a residue not exceeding 20 per cent. The coarse sieve may be 76·2 gauge, or 5806 meshes, to correspond with the 900 per square centimetre sieve.

A less severe test is to pass through a sieve of 22,500 meshes (150 gauge), with a residue of 25 per cent.; or to pass through a 10,000 meshes sieve (100 gauge), with 12 per cent. residue.

Mr. Carey suggests a residue of 8 per cent. on a 22,500 meshes sieve (150 gauge), and 30 per cent. on a 32,400 meshes (180 gauge sieve). Mr. Watson, of Messrs. Johnson & Co., Newcastle-on-Tyne, recommends a residue of 19 per cent. on 10,000 meshes, and 31 per cent. on 32,400 meshes; also an 8 per cent. residue on 5806 meshes (76·2 gauge). Dr. Michaelis gave for a 14,400 meshes sieve, a residue of 33 per cent.; this is too coarse.

It may be accepted that coarse particles, rejected by a sieve of 6400 meshes, have little or no cementitious value, and may be considered as an adulteration of the cement by so much sand. Hence the low sieve test should be absolute.

As regards the high sieve test, the finer the cement powder, the stronger the cement, the larger area of

TEST FOR FINENESS OF GRINDING. 91

enveloping film it will make, and the more sand, or broken stone, can be mixed with it while maintaining the same strength as is possessed by the same cement when coarsely ground. A high sieve test therefore assists an engineer in procuring a superior quality of cement; alone, it is insufficient, as an underburned cement can be easily ground to a fine state of division.

Sometimes an allowance is made if the cement does not attain to the high standard on the 32,400 meshes sieve. It may be accepted with a residue of 30 per cent., provided that there be an increase of 10 per cent. made in the proportion of the cement used for concrete and mortar mixing, and no extra charge is to be made for the additional quantity of cement used over and above the quantity specified to be used for the particular work or purpose. With reference to the higher sieve tests, it has been shown by experiment that cement powder passing a No. 103 gauge sieve possesses materially less strength than the same cement after being sieved through a 176 gauge sieve.

78. French and German manufacturers and engineers use sieves of 900 meshes per square centimetre (5806 meshes per square inch, or 76·2 gauge), and 5000 meshes per square centimetre (32,258 meshes per square inch). The English 180 gauge sieve corresponds with sufficient accuracy to the latter. They may be taken as the maximum and minimum mesh used at present.

79. In specifying the number of meshes to the square inch of surface of sieve, it is equally necessary to

define the thickness of the wire or fibre of which the sieve is made. Sieves are made of brass wire, and the finer meshed are sometimes of silk, but the finest sieves can now be made of woven wire cloth.

As a general rule, the thickness of the wire is not more than one-half the side of a mesh opening, and for a 50 gauge sieve (2500 meshes), the wire is No. 34 B.W.G. For the Vyrnwy reservoir works, a 60 gauge sieve was specified to be made of brass wire gauze weighing $3\frac{3}{4}$ oz. to the square foot. A silk sieve of 176 gauge (30,976 meshes) gave a mesh opening of about 0·004, and a diameter of fibre of about 0·00168 of an inch. A silk sieve of 103 gauge gives a diameter of fibre of about 0·0022, and a mesh opening of about 0·0075 of an inch. A wire sieve of 5806 meshes will have a wire of about No. 36 B.W.G. (or about 0·0045 diameter).

Prof. Baker gives in an American (U.S.) specification, a No. 50 gauge sieve (2500 meshes), the wire is to be No. 35 Stubs' wire gauge; a No. 74 gauge sieve (5476 meshes), wire to be No. 37 Stubs' gauge, No. 100 gauge sieve (10,000 meshes), wire to be No. 40 Stubs' gauge.

Sand sieves to be No. 20 (400 meshes), wire to be No. 28 Stubs' gauge; and No. 30 gauge sieve (900 meshes); No. 31 Stubs' wire gauge. Stubs' gauge appears to be similar to the B.W.G.

The sieves are furnished in sets, made of best quality of brass wire cloth set in metal frames, cloth to be true to count of meshes and wire of the specified gauge. The sieves are nested, No. 100 being 7 inches diameter; No. 74, $6\frac{1}{2}$ inches; and No. 50, of

SPECIFIC GRAVITY OR DENSITY TEST.

6 inches diameter. The sand sieves to be 8 and $7\frac{1}{2}$ inches diameter respectively. These sieves are made by Williams, of Fulton Street, New York. It is stated that in ordinary fine sieves, the number of wires may be about 10 per cent. less than the stated number of the sieve gauge. For accurate and comparable results the sieves should be accurately made, and the fixed proportion of diameter of wire to be equal to the half of the side of a mesh should be adopted for fine meshed sieves.

The German sieves appear to be now made to the rule that the diameter of wire is to be one-half the side of the mesh opening.

SPECIFIC GRAVITY OR DENSITY TEST.

80. The specific gravity of the cement to be ascertained by an approved method, and to be not less than 3·1 for recently made cement dried in a water-oven for 15 minutes at a temperature of 212° F. For a cement more than three months old the specific gravity may be lowered to 3·07.

81. The cement when tested should be carefully observed to ascertain (1) the time taken to set hard, using Vicat's needle test or similar test (see Art. 97); (2) the temperature of the cement at the moment of complete mixing, and at regular intervals during the time occupied in setting. A material increase of temperature is an unfavourable sign.

Pats of neat cement, 3 inches diameter, $\frac{1}{2}$ inch thick at centre, and $\frac{1}{4}$ inch thick at edges, may be used for these tests. Some pats to be immersed in cool water one hour after mixing, and in 48 hours must show no signs of cracking, or of disintegration. Hot water immersion is given in Art. 87.

82. In Germany the Cement Manufacturers' Association has framed, by agreement, the following specification of tests:—

The test of tensile resistance of briquettes made of cement and normal sand in the proportion of 1 to 3 by weight. The normal sand is specially prepared and is supplied by a Governmental department, the size of the grains being regulated by a sieve test, nearly similar to the English sand. The briquettes are moulded under similar conditions, and are immersed, one day after moulding, for 27 days in water.

The tensile resistance of the mortar briquette, 28 days after moulding, 27 in water, is to be not less than $227\frac{1}{2}$ lbs. on the square inch on a section of 0.775 of an inch. Some German makers are prepared to make a cement to bear a test of 300 lbs., and up to 350 lbs. on the square inch under the same conditions.

The Society of Austrian Engineers and Architects specify a seven day test of the mortar briquette of 114 lbs. to the square inch, and at 28 days, 171 lbs. nearly.

The German sieve test for fineness of grinding is that the cement must pass through a sieve of 5806 meshes to the square inch, made with brass wire of a fineness of gauge of about 0.0045 inch, or thickness equal one-half the mesh width. The residue of coarse particles left on the sieve must not exceed 10 per cent. of the quantity tested.

Some of the German manufacturers are prepared to

CEMENT TESTING (AMERICAN PRACTICE). 95

supply a cement that will pass a sieve test of 14,400 meshes, and even of 32,000 meshes to the square inch, with a residue not exceeding 10 per cent. Recent German rules for testing are given in Proc. Inst. C.E., vol. xci. p. 474.

A French specification gives, for tensile resistance of 0·775 square inch section mortar (1 to 3) briquettes, at 7 days, 114 lbs. per square inch ; and at 28 days, 213 lbs. per square inch. The increase in resistance from 7 days to 28 days to be at least 25 per cent. The specification for the Boulogne Harbour Works will be found, in a summary, in Proc. Inst. C.E., vol. lxxxviii. p. 457.

CEMENT TESTING (AMERICAN PRACTICE).

83. Samples of cement for testing are to be taken from the interior of the original packages as delivered by the maker, from one out of every five, or from every ten packages containing about 20 bushels of cement. The samples to be stored in air-tight boxes till used.

The sand for making the mortar briquettes is to be from crushed white quartz rock, or from a hard quartzite (sand similar to that used in making sand-paper), and is to be sieved to pass through a 400 mesh, and to be rejected by a 900 mesh, per square inch brass wire sieve. Sand to be weighed out in the proportion of 3 to 1 of cement, and to be first mixed dry with the cement, then water at about 65° F. is to be added, and the mixing completed forthwith. The amount of water to be used is 10 per cent. of the combined weight of the sand and cement. The briquettes to be of the standard size, having a sectional area of one square inch at the expected point of rupture; to be moulded in gun-metal moulds.

The testing of the tensile resistance of the briquettes to be carried out with an approved lever machine, and the loading is to begin with one-half of the specified resistance, and the stress is to be increased regularly and uniformly at the rate of 400 lbs. per minute. The briquettes are to be tested 7 days and 28 days after moulding for neat cement (being immersed in water for 6 days and 27 days respectively); and at 28 days for mortar briquettes, after 27 days in water.

The neat cement briquettes are to carry at least 250 lbs. and 550 lbs., at 7 and 28 days; and the mortar briquettes at least 150 lbs. at 28 days (for 1 to 3 sand briquettes).

The time of setting to be noted by the vertical wire test (Gillmore's); a $\frac{1}{12}$-inch diameter wire, with flat end, to be loaded with $\frac{1}{4}$ lb., and a $\frac{1}{24}$-inch diameter wire to be loaded with 1 lb.; to be tried on cakes of neat cement, 2 to 3 inches diameter and $\frac{1}{2}$ inch thick, edges $\frac{1}{4}$ inch thick, one cake to be in air, and one immersed in water at about 60° F. The time at which the loaded wires cease to penetrate into the pat is to be noted.

The sieve test is to measure the percentage, by weight, of cement from each sample rejected by sieves of 2500 (50 gauge), 5476 (74 gauge), and 10,000 meshes (100 gauge) to the square inch, the gauges of the wire forming the sieves being Nos. 35, 37 and

40 of Stubs' gauge respectively. The percentage rejected from the three sieves to be zero (?), 5 (?) per cent., and 15 (?) per cent. respectively. [The amount of residue is not definitely stated, the values inserted are for a high class cement.]

The bulk of a given weight of cement, say 5 ozs., to be ascertained by filling the cement powder into an upright cylinder 6 inches high, and having a sectional area of 2 square inches. On to the top of the powder is to be lowered a frictionless piston weighing with attachments 50 lbs. The load to remain on the cement for an hour, and then the bulk of the cement is to be accurately estimated.

Simple Tests without the use of a Tensile Testing Machine and Moulded Briquettes.

Adhesion Test and Expansion Test.

84. Engineers often require small quantities of cement, and also are liable to be users of cement under circumstances preventing ready access to testing machines. In such case an adhesion test, and a simple test for expansion and contraction of the cement during setting and hardening, may be adopted with satisfactory assurance of detecting a faulty cement, though the results of the tests cannot be definitely compared with those of other cements, excepting in the case of adhesion.

Two rectangular pieces of limestone, or sandstone, are to be cut with parallel faces, in shape like an ordinary brick, and 4 in. wide by 9 in. long and 3 in.

H

thick; or sandy bricks rubbed down to smooth surfaces and right-angled edges; stones and bricks are to be washed clean and are to be soaked in water. The cement is to be mixed to a stiff paste, either neat, or with 3 of sand to 1 of cement by weight, and the two stones or bricks cemented together crosswise at right angles.

The cemented faces of the stones or bricks must be plane, but not too smooth, generally the washing and scrubbing, with a soft brush, of the plane surface will produce sufficient roughness. The cemented blocks must be kept in a damp cloth till tested. After seven days being allowed for setting and hardening of the neat cement, the test of adhesive strength is carried out by placing the upper block on firm supports, and suspending, or loading weights symmetrically on the lower block until the mortar joint gives way, which should not occur (for neat cement) under 90 lbs. on the square inch.

If the means of loading are scanty, smaller dimensions of blocks may be adopted, down to $1\frac{1}{2}$ inch by 1 inch broad, and $\frac{3}{8}$ to $\frac{1}{2}$ inch thick.

With a cement mortar composed of 2 of sand to 1 of cement, the adhesion of bricks, having a cemented surface of about 40 square inches, should resist a stress of from 15 to 30 lbs. on the square inch at 28 days after cementing.

Portland cement adheres to some hard surfaces, such as granite and ground plate glass, much better than to some softer materials, such as brick, Portland limestone, &c., but it does not adhere well to a wood surface.

85. *Adhesion Test.*—Mr. Mann has designed a simple form of apparatus for carrying out his system of testing the adhesion of Portland cement. (See Proc. Inst. C.E., vol. lxxi. p. 251.)

Two pieces of limestone are sawn out of solid stone, they each measure $1\frac{1}{2}$ inch by 1 inch, and are $\frac{1}{4}$ to $\frac{3}{8}$ inch thick. They are then cemented together (by the neat cement to be tested) in a cruciform shape. Ground plate glass, granite, or sandstone (fine grained) may also be used. Porous stones must be saturated with water before they are cemented together.

The cruciform block is kept in air for 7 or 28 days, or in water for 7 or 28 days after cementing together, and is then tested in a lever testing machine.

For testing, the cruciform block is fitted in the machine by placing one bar (the upper) resting on two supports under the end of the short arm of a lever. On the transverse lower bar of the block is placed a forked or U-shape of steel, and the broadened out ends of the limbs of the inverted U rest on each projecting end of the lower bar of the block. On the top of the inverted steel U is a hardened conical point, on which the end of the lever rests.

With a finely-ground cement of good quality the adhesion should be about 90 to 100 lbs. per square inch of cemented surface, after seven days in water. The nature of the surface is, however, of importance; it should be perfectly clean, coarse in texture, and compact.

The test shows conclusively the superiority of finely ground over coarsely ground cement. A table of experiments is given by Mr. Mann, in Proc. Inst. C.E., vol. lxxi. p. 253.

Ordinary coarsely ground cement gave an adhesion ranging from 41 lbs. to 76 lbs. per square inch of surface, at 7 days. The cemented blocks were immersed in water immediately after being joined together with neat cement.

86. A simple test suggested and used by Mr. Charles Stevenson is to mould (two or three) bars of neat cement, 8 or 12 inches long, and exactly 1 inch square in section. For constant use, the mould is made of gun-metal; but one could be easily made of planed wood, or of sheet metal, to be washed over with soap solution on the mould faces.

The bar, in not less than two hours or more than ten, is immersed in water (at about 60° F.), and kept there till 7 days after moulding, when it is broken by loading it at mid-length when placed symmetrically on rigid supports 6 inches apart; the breaking load to be not less than 75 lbs. The weight is added gradually and without shock, and is to be suspended from an iron or steel bar, $\frac{1}{2}$ inch broad, with rounded surface (struck to a radius of $\frac{1}{2}$ inch) resting on upper surface of the cement bar at midway between the supports. (The weighting may begin with 40 lbs., with regular increments of 5 lbs. up to 60 lbs.; 2 lbs. up to 70 lbs., and then 1 lb. at a time till the bar breaks; weights to be added at ten seconds' intervals.)

87. Simple tests of cement pats in hot water. Neat cement is to be gauged with the usual quantity of water, and be moulded by hand into small circular

slabs about 3 inches diameter, $\frac{1}{2}$ an inch to $\frac{3}{4}$ inch thick at the centre, and $\frac{1}{4}$ inch thick at the edge. At (one to three) hours after setting, each pat to be immersed in pure water maintained at 212° F. After from 3 to 7 days' boiling, the pats are not to show any sign of cracking or of disintegration. This boiling test appears to be an excellent safeguard against the use of a treacherous underburned cement, as in Proc. Inst. C.E., vol. cvii. p. 150, will be found two analyses of cements almost identical in composition, but the one withstood perfectly the boiling water test, while the other, probably much underburned, was disintegrated in 3 hours.

Another rapid test is to make small cakes of neat cement 5 to 6 inches diameter, or square, and $\frac{3}{4}$ inch thick at centre, and $\frac{1}{4}$ inch at edges, to be placed in water as soon as the cement has set; they must not be soft nor cracked at the edges in 24 hours. Then they are to be heated up to 212° F., and while hot are wetted with cold water; there must be no disintegration. An additional test is recommended, that of desiccating at 212° F. a boiled briquette. (Mr. Dyce-Cay.) Another pat test is to make a pat 3 to 4 inches square and $\frac{1}{8}$ inch thick, on glass, to be immersed in water as soon as mixed. The cement is to be hard in 6 hours, and the pat is not to turn yellow on surface, nor to crack during 7 days' immersion.

88. *Contraction Test.*—A test for expansion and contraction of Portland cement, while setting and hardening, is to fill a thin clear glass bottle with mixed neat cement and allow it to set and harden; if any appreciable expansion occurs the bottle will be

cracked, and if there be contraction it will be possible to pour a coloured liquid into the space between the cement and the sides of the bottle. Expansion may be also noted by any increase of the height of the cement in the bottle. A swelled glass lamp chimney is recommended as excellent for the purpose.

The Prussian test for constancy of volume of a cement while setting and hardening is stated to be— neat cement mixed with water to the consistency of thick cream, and poured in a thin cake on glass or any impervious slab. After setting, cake and slab to be placed in water, and there is to be no crumbling or cracking at the edge in one or more days. A similar cement cake is also baked for an hour on a heated iron plate, and must not disintegrate nor crack at the edge. Also a similar cake, after 23 hours in air, is placed in boiling water for an hour; it must not crack nor fly to pieces.

ADULTERATION OF CEMENT.

89. Cements are sometimes adulterated by the admixture of blast furnace slag with the broken clinker before grinding, also furnace hearth scoriæ. Sometimes light-coloured overclayed or underburned cement powder is darkened by the addition of lampblack, or similar substance. The slag addition gives a greenish tint, and also greater density, the appearance of high burning. The German Union of Cement Makers decided that all mixtures of solid substances with Portland cement were to be regarded

as adulterations, an exception being made in favour of the addition of not more than 2 per cent. of plaster of Paris, which is considered to improve the cement. It is on the whole advisable to consider every addition to the pure cement as an adulteration; the engineer using the cement can, if he wishes, add plaster of Paris, but there is risk of lessening the durability of the cement by the addition of sulphates.

Adulteration by blast furnace slag, or by colouring matter, may be detected by throwing a handful of cement powder into a soup plate full of clean water, the colouring matter will float. Stir the cement powder in the water till it becomes thoroughly mixed, pour off slowly the liquid mixture, the slag will sink to the bottom, and remain in the plate when the muddy liquid has been poured away.

Slag adulteration is neutralised to some extent if there be an excess of lime in the cement powder.

In a case of adulteration of cement the following observations were noted :—The cement mortar was made of 1 cement to 1 sand, and in mixing, the plasticity or fatness of the mortar was noticeable ; on the surface appeared a soft white film, which, when dry, became yellow, and was easily wiped off. A thin pat of the mortar, dried in air, became fissured with cracks. On some of the cement being thrown into water, about 1·2 to 1·5 per cent. floated on the surface, and carbonaceous particles were detected. On treatment with hydrochloric acid, sulphuretted hydrogen was evolved, and there was a considerable residue of insoluble matter. The cement may have been adulterated with blast furnace slag, or may have been made of inferior materials badly mixed or improperly burned.

A report of Messrs. Fresenius on the adulteration of Portland cement is summarised in Proc. Inst. C.E., vol. lxxix. p. 377.

Acceptance and Storage of Cement. Removal of Rejected Cement.

89A. On acceptance of the cement from the makers, some engineers adopt the practice of emptying the bags or barrels on to the floor, or on to shelves or trays fixed in the weather-proof shed, to secure proper aëration, the cement powder being spread in a layer not exceeding 4 inches thick, and not more than three such layers to be allowed. Some increase of bulk in the cement powder may occur, ranging between 5 and 10 per cent.

The floor of the shed should be capable of containing at least (7) days' maximum consumption of cement, the total depth not exceeding 12 inches (or 24 inches).

All cement rejected by the engineer is to be removed, in not more than (3) days' time after notice of rejection has been given, at the sole expense of, and by, the maker. If the maker fails to remove rejected cement, the engineer will remove it, and will deduct the cost of such removal from the sum of money payable in respect of accepted cement. The engineer will not be answerable in any way for the preservation of rejected cement, even when removed by him.

CHAPTER V.

WEIGHT TEST; FINENESS OF GRINDING; TENSILE STRESS TEST; STANDARD SAND; DENSITY TEST; CHEMICAL ANALYSIS; VICAT'S NEEDLE TEST; RE-BURNING OF DAMAGED CEMENT.

CONSIDERATION OF TESTS OF PORTLAND CEMENT.

90. *Weight, and Fineness of Grinding.*—The weight test imposed by some engineers, viz. that the cement is to be lightly filled (by sliding down an inclined surface at an angle of 45°) into a bushel measure, and that the striked imperial bushel of cement so filled is to weigh from 112 to 118 lbs., is now generally recognised to be useless, if not even misleading; the weight alone of the cement is no indication of its quality. Mr. Faija recommends a shoot inclined at 35° to the horizontal, and the delivery from shoot to be at 5 inches above edge of bushel measure. A coarsely ground cement will weigh heavier than a finely ground, and a very finely ground cement will weigh only about 90 to 96 lbs. per striked bushel. The imposition therefore of a heavy weight test is harmful, as it tends to encourage the maker to grind coarsely, or to produce a soft underburned easily ground clinker, whereas the only rational system is to use a well-burned cement in the finest possible state of division. If therefore a weight test be specified, it should be low, and should be accompanied by a stringent sieve test.

Coarse particles of cement, rejected by a 5806 mesh sieve, appear to have little or no cementitious value, and may be regarded as an adulteration with so much sand; the coarse particles remain inert, and are but slightly changed on their surface.

The weight specified for finely ground cement should never be much higher than 100 lbs. per imperial striked bushel. A cement, ground in the ordinary coarse fashion to a weight of about 115 lbs. per bushel, was reground to pass through a 32,000 mesh sieve without residue; it then weighed only about 90 lbs., and mortar briquettes made of this finely ground cement had, at 7 and 28 days respectively, from 3 to $2\frac{1}{2}$ times the strength specified in the rules of the Society of Austrian Engineers, &c. Dr. Michaelis considered that a high class finely ground cement should not weigh more than 75 lbs. per cubic foot or $96\frac{1}{4}$ lbs. per striked bushel.

An over-limed cement, highly burnt and coarsely ground, would probably come up to the ordinary weight test of 115 lbs., but the cement powder would be of little strength, and possibly inert. As an instance of ordinary practice, the weight of a cement, ground to pass a sieve of 5800 meshes per square inch with 8 per cent. residue, is specified to be $112\frac{1}{2}$ lbs. per striked bushel.

As the method of filling the bushel measure will exercise an important influence on the quantity of cement that can be packed into it, there is a paramount necessity for an engineer adopting a weight test, to specify very exactly the method of filling, and also to ensure the method being accurately carried out.

An experiment cited by Mr. Grant gives marked testimony in favour of fineness of grinding. Measured quantities of standard sand and ordinary cement were taken in the usual proportion of 3 to 1. The measured quantity of cement was then sieved to 5806 meshes per square inch, and the residue, ranging from 5 to 15 per cent., was thrown away. The sifted cement (85 to 95 per cent. only of the measured quantity) was made into mortar briquettes with the full original proportion of sand, and, when tested, three briquettes broke with a higher tensile stress than similar briquettes made of the unsifted cement in full quantity.

Prof. Hayter Lewis, on examining with a microscope Portland cement mortar gauged with fine sand, observed that each particle of sand appeared to be enveloped with a film of cement, and it was the union of these films at points of contact that gave cohesion; there were vacuities between the grains of sand. If the cement be ground as fine as possible, a given quantity will make more perfect film, and a greater area of surface will be covered, than with a similar amount of coarsely ground cement.

The extra expense incurred by fine grinding may be partly compensated by the ability of fine cement to carry a larger proportion of sand than, and retain the same standard of strength as, a coarse cement.

91. *Testing by Tensile Stress.*—Tensile stress is used to pull asunder a briquette of neat cement, or of cement mortar, because it is easily applied, and the machinery required is more simple and less costly than that for testing under compressive stress, though the latter would seem to be more appropriate.

There appears to be no definite relation between the resistance of Portland cement to tensile and compressive stresses. Messrs. Dyckerhoff give 20 as the ratio, while Prof. Bauschinger states that compressive resistance may vary from eleven times to seven times the tensile resistance.

Formerly the test briquettes were made solely of neat cement; of late years a sand mortar briquette has been added, and the test result of this is the more important of the two. The admixture of sand with the cement lessens its strength, and also retards the process of setting and hardening, but on the other hand it develops a very important quality in the cement, that of adhesion, and it is only when mixed with sand that the great superiority of a finely ground over a coarse cement is clearly manifested. It appears that there may be but little difference between the tensile resistance of a fine and coarse cement when the briquettes are made of cement only; in some cases the coarse cement may seem to be the stronger. When mixed with the usual proportion of three times its weight or volume of standard sand, the superiority of the fine cement is strongly marked, and the sand mixture approaches more closely to the actual conditions of use in engineering structures. Dr. Michaelis states there is no better test for the detection of faulty cements than the strength of a cement and sand mixture.

92. An objection to the use of sand is that the strength of the briquette largely depends upon the size of the grains and upon the coarseness of texture of their surface; and that there is difficulty in procuring

sand of such uniformity of quality as will permit of close comparison between any different series of testings. Another objection is to the time that must necessarily lapse between the delivery of the cement and its acceptance after satisfying the tests. With neat cement the decision used to be given in 7 or 8 days, with the sand mixture 28 days is the minimum time required. On works where large quantities of cement are used, at least five weeks' supply must be in store when a new delivery is received.

Standard sand of uniform quality is best obtained by crushing quartz, or a quartzite rock of suitable durable quality; the sand thus obtained is sieved as before stated (Art. 72) to reject coarse and fine particles, and the selected grains must be washed clean from any dust. The sand should be prepared at one establishment for the supply of a country or a district of convenient size, and the make of many days' work should be thoroughly mixed together; this establishment should be the sole source of supply for the district.

Waterworn sands are of such uneven coarseness of texture that tests made with them are not comparable. Standard sand, when properly prepared, gives satisfactory results.

The lapse of time occasioned by the use of mortar briquettes can be now shortened by the immersion of the test briquettes in water at 212° F., or at 170° or 180° F. if preferred; it appears that 6 days' immersion in boiling water may be substituted for 27 days in cold water.

The use of water at 212° F. for mixing the cement is

recommended as a method of detecting excess of lime; it is stated that if the lime is more than 64 per cent. the cement will not harden, but probably the extent of calcination has an important influence.

93. It is not advisable to specify a high resistance to tensile stress at 7 days, as it leads to overliming of the cement; about 350 to 400 lbs. on the square inch sectional area of briquette of neat cement is sufficient. There should be a steady increase of resistance at the 14 and 28 days' test.

The amount of water used in mixing should be ample for the perfect hydration of the cement, and the correct proportion is best ascertained by experiment with the cement. It is better to use a slight excess of water rather than too little, the cement, in setting, will reject any surplus; it is very important to prevent loss of water of hydration from the briquette.

94. *Density or Specific Gravity of Cement.*—The density of cement is fairly constant and not dependent on the size of the ground particles, but varies with the proportions of lime and clay, with the degree of burning, and with the amount of exposure to air that the cement has undergone. A newly ground cement of good quality should have a specific gravity of from 3·1 to 3·5, and a cement of less than 3·1 should be suspected of imperfection; it is generally a sign of deterioration of the cement from over long keeping, or of underburning of the clinker. A low specific gravity with a high tensile resistance at 7 days is a sign of an untrustworthy cement that will probably become weaker, and even disintegrate, with age.

Deterioration in density increases with age, the rate

of diminution is stated to be about 0·13 in four months, but the rate depends upon the conditions attached to the keeping of the cement. The specific gravity of the cement should not be less than 3·1 within 1 month after manufacture, and not less than 3·07 at 4 months. Mr. Mann describes a simple means of ascertaining the density of cement in Proc. Inst. C.E., vol. xlvii.

In Schumann's method of determining specific gravity, the displacement of turpentine spirit in a graduated glass tube, by the insertion of a measured quantity of cement, is the method adopted. The cement is thoroughly dried at a temperature of 212° F. for at least 15 minutes.

The turpentine spirit must be free from all moisture; this may be ensured by mixing with the turpentine some fresh cement before the test, and decanting off the clear liquid, the cement will absorb all the moisture present.

A description of the determination of the specific gravity of a cement is given in Proc. Inst. C.E., vol. lxxix. p. 377, in a summary of a report by Messrs. Fresenius; also in vol. cvii. p. 46.

95. *Chemical Analysis.*—A chemical analysis of a cement powder cannot be depended upon as evidence of its quality as a cementing substance, inasmuch as the analysis usually furnished does not distinguish between a badly mixed and an imperfectly chemically combined cement, and one that has been properly made in every stage of the process, and, above all, it does not show whether the cement clinker has been well- or ill-burned.

An analysis should show in separate values the free and the combined lime, the insoluble and the soluble silica, otherwise it is of little use even in assisting the investigation of cause of cement failures. There is, however, a practical difficulty preventing the distinction in any analysis of free from properly combined lime.

A small amount of sulphuric acid is always present in coke-fired kiln-burned cement. If the proportion be larger than $1\frac{1}{2}$ per cent. it is injurious; in fact the presence of any acid in the cement is a source of weakness, and an excess of soluble salts leads to slow destruction of cementitious properties. The maximum of sulphate of lime should be about 1 per cent. Iron oxide may be present in the cement up to the proportion of about 3 per cent.

As the part played by the alumina of the clay does not appear to be definitely settled, it is difficult to specify any exact proportion for a high quality of cement. Knapp appears to regard alumina and iron oxides only as components of double silicates, assisting the formation of silicates of lime. The best cements appear to contain from 5 to 9 per cent. of alumina; on the other hand the cement made at Skinningrove from blast furnace slag contains by analysis about 25 per cent. of alumina, and according to the experience of about 5 years this cement makes an excellent durable concrete in sea water. (See Art. 26.)

One authority specifies that a cement in which the total of the combined silica and alumina is less than 44 per cent. of the lime, is to be looked upon as of doubtful quality. The silica in high class cement is generally not less than 22 per cent.

96. A Portland cement is considered to be quick setting if the neat cement sets in air in 10 minutes after mixing with water; and to be slow setting if the time occupied be 30 minutes and upwards, to 5 hours as a limit. Failure to set in 5 hours may entail rejection. The setting can be tested by placing a weighted flat-ended steel needle vertically on the surface of the pat of cement, or on a briquette in the mould; the beginning of setting is marked by the sticking of the needle in the cement. The needle may be $\frac{1}{12}$-inch diameter, and be weighted to 4 ozs. total weight; when the setting is complete the needle fails to penetrate the surface. Mr. Mann recommends a vertical steel needle $\frac{1}{16}$ inch diameter, flat ended, moving freely in guides, and loaded to weigh 1 lb. "Setting" to be when the needle makes no visible mark on the surface of the pat. Some experiments on fine and ordinary cements gave for "fine," setting in air in 10 minutes to 5 hours, and for "ordinary," 40 minutes to 8 hours, at 7 days after the grinding of the cement; the times are for neat cement. Setting in air appears to be much quicker than in water, but there is no definite ratio. The age of the cement after being ground exercises a material influence, an old cement being slow setting.

97. Vicat's needle test, as used at the Boulogne Harbour works, is stated to be $10\frac{1}{2}$ ozs. (300 grammes), attached to the needle of 1 square millimetre (0·00155 square inch). At Dunkirk the load was 1·53 kilo. (3·37 lbs.). At Boulogne a slow setting cement

seems to have been used, as the specification states that a pat of cement mixed with sea water at 59° F. (15° C.), and supporting the above initial needle test in less than half an hour was to be rejected, and the final setting to be in not less than 3 hours. The tensile resistance of this cement at 7 days (6 in water) to be 178 lbs. per square inch. Initial setting is when the needle fails to penetrate to the whole depth of the pat (1·575 inches); complete setting when the surface of the pat supports the needle.

98. The German needle test is stated as follows: a needle weighted to 10 ozs. and 0·0015 square inch in sectional area, is placed vertically on the top of a pat of mixed cement paste. The needle must pass freely through guides, and be flat ended. The cement pat may be 1½ inch thick and may fill the interior of a rectangular metal ring or cylindrical box 3·15 inch interior diameter. The point of complete setting is the moment when the needle fails to impress the surface of the cake; the beginning of the setting when the needle fails to penetrate completely through the pat; the time taken in setting is the interval between these two results. A needle test with adjustable weight is mentioned in Proc. Inst. C.E., vol. lxxxix. p. 501.

99. *The Re-burning of Damaged Cement.*—A cement, from neglect or accident, may have been exposed to access of dampness, or of moisture, and in a country distant from the place of manufacture it is an important question for an engineer whether such cement can be made fit for use by re-burning. Dr. E. Nicholson, late Army Medical Department, states

that damaged cement re-burnt may be used, but that it displays great energy in setting and hardening, setting in about 2 minutes under water; and therefore it is advisable to mix with it about $1\frac{1}{2}$ to 2 times its bulk of ordinary cement, for setting in about 10 to 30 minutes.

The re-burnt cement appears to show better adhesion (after 18 hours' immersion) than ordinary cement.

The damaged cement, ground to a fine powder, is made into a paste with water, and is spread to air-dry in a layer about 2 inches thick on a stone floor; the layer to be marked off while soft into convenient sized bricks, say 6 inches square; after drying for 24 hours the bricks may be stacked till wanted for burning.

Burning is carried out in a kiln, and 6 to 8 hours' calcination is stated to be sufficient; also that the temperature may be below that required for burning lime, but this point should be carefully investigated. Probably the best temperature for re-burning will be somewhat below that of the original burning.

After burning, the bricks are broken up and are ground to the finest possible powder. The powder is aërated and cooled for 48 hours after grinding, and is then packed in barrels, which are to be as air-tight as possible, and are to be stored in a cool dry place. It is not advisable to keep the re-burned cement for more than 3 months before use.

Re-burning caked Portland cement, without crushing and mixing with water to a paste, does not appear to give a successful result.

CHAPTER VI.

SLAKING OF LIMES; QUALITY AND PROPORTION OF SAND AND WATER; HAND AND MACHINE MIXING; CEMENT MORTAR.

LIME AND CEMENT MORTAR.

100. MORTAR is made of powdered calcined lime, sand, and water in definite proportions. Calcined limestone is to be purchased either in the lump just as it comes from the kiln, or the lumps have been crushed and ground to a fine powder. Unless implicit reliance is to be placed upon the quality of the powder, it is better to purchase the calcined lumps, and reduce them to powder under inspection when required for use.

Ordinary pure or chalk lime in calcined lumps can be slaked, that is reduced to hydrate of lime in the form of a dry powder, by sprinkling the lumps of quicklime with water in the proportion of about three to six gallons to the bushel of lime. Pure lime is often slaked by total immersion in water for some days. Gray-chalk lime is slaked by throwing water over the quicklime lumps, and covering them with sand to retain heat.

Blue lias and other limes possessing a high degree of hydraulicity do not slake readily. Some hydraulic limes will slake if, after being wetted, the heap of watered lime is covered with sand, and with sacks, &c., to retain heat in the mass. Slaked blue lias lime should always be sifted before use, so that all the imperfectly slaked pieces may be separated and again treated.

As it is essential that hydraulic limes be in a finely divided state before being mixed into mortar, and as they are all more or less inert and slow in slaking, it is better to grind the lumps of calcined lime to a fine powder before the water is added, and thus ensure the complete slaking of every particle. If mortar made of imperfectly slaked lime be used in masonry, the unslaked particles begin to slake and expand in the mortar joints, and brickwork may be ruptured from this cause. As an example of the force exerted by the expansion of lime during slaking may be cited the modern use of cartridges of compressed quicklime as a substitute for blasting powder, in dislodging coal in dangerous fiery seams.

Ordinary blue lias lime in the lump requires about $2\frac{1}{2}$ gallons of water per bushel for slaking, and about $4\frac{1}{2}$ gallons for mixing into mortar; if ground to fine powder, about 5 to $7\frac{1}{2}$ gallons per bushel for mixing. In slaking hydraulic limes it is very important to prevent loss of heat, and ample time should be allowed, two days at least for moderately hydraulic limes, up to a week for superior limes.

101. The proportion of sand mixed with the slaked or ground lime is regulated by the quality of the lime and the purpose for which the mortar is to be used. A larger proportion of sand may be mixed with rich chalk lime, without seriously diminishing its low cohesive and adhesive strength, than can be used with a hydraulic lime produced from blue lias limestone.

Sand is mixed with lime for three reasons :—(1) To lessen the quantity of lime used to make a given bulk of mortar; sand is generally cheaper than lime, and the cost of the mortar per cubic yard is thus reduced. (2) To confer on the mortar somewhat greater resistance to crushing, and also to prevent excessive shrinkage during the setting and hardening of the mortar in use. Mortar made from lime alone contracts when hardening. (3) To separate the particles of lime and render the mortar more porous, thus with the purer limes facilitating the penetration of carbonic acid, and accelerating the setting of the mortar. It is considered also that the crystallised particles of carbonate of lime adhere more firmly to the sand grains than to each other.

It must be remembered that the lime forms a thin film round the grains of sand, and tends, more or less, to fill up the voids between the grains; consequently the volume of mixed mortar is considerably less than the sum of the volumes of the separate ingredients. A deduction of about one-fourth is generally made, that is, the volume of mortar equals three-fourths of the sum of the volumes of sand and lime. But this deduction varies with the quality of the sand, and is least with the finest grained.

Quality of Sand.

102. Sand should be in all cases clean, free from clayey matter or vegetable earth; when rubbed between the hands should not soil them, and when dropped into water should not cause muddiness;

should consist of sharp angular *siliceous* fragments not less than $\frac{1}{24}$-inch in diameter, and not exceeding $\frac{1}{8}$-inch; and should have a rough texture of surface. Calcareous or argillaceous sands are unsuitable, as the former will dissolve, and the latter partly dissolve in acids.

Fine sand may be defined as composed of fragments whose diameters range between $\frac{1}{24}$ and $\frac{1}{16}$ of an inch, coarse sand between $\frac{1}{16}$ and $\frac{1}{8}$ of an inch.

For making béton Coignet, or sand concrete, a sand having grains $\frac{1}{3}$-inch diameter, about the size of a small bean or of a pea, may be used.

The strength of mortar is found to depend largely upon the size of the grains of sand, and upon the coarseness of texture of the grains. In sands of equal size of grain, the coarsest texture is the best; and in sands of equal coarseness of texture, the larger grained are the best. Coarseness of texture is, however, of greater importance than size of grains, and sand grains of uniform size throughout do not necessarily make the best mortar.

Sands that are much waterworn, and consist mainly of rounded grains, are not so suitable for making either mortar or concrete; they are, however, often used in default of a better material.

The colour of sands is very variable, and depends upon the metallic oxides present with the silica. The best siliceous sands have been produced *in sitû* by the disintegration of granitic, or of granulated quartz, rocks, and have not been waterworn, they are sometimes styled "virgin" sands.

103. There are two descriptions of sand in general use, "river" and "pit" sand. "River" sands are

dredged up from the beds of rivers, either directly from sand-banks, or mixed with large coarse gravel, from which the sand is separated by washing and sifting. They are liable to be waterworn, and also to be mixed with mud and organic matter, thorough washing in clean water is essential.

104. "Pit" sands are found in layers or pockets among the rocks and soils of the earth's crust. These sands have all been deposited in place by the action of water at some period more or less remote. The grains are generally waterworn and rounded, and sometimes too small to rank among the best class of sands; and may be loamy, that is, coated and mixed with clayey matter.

105. "Blown sand" is the fine-grained sand found forming hillocks and ridges at many places on the sea coast. This sand is generally so fine grained that it is swept up and borne along by a high wind; the grains are much rounded, and have a comparatively smooth surface. To made mortar or concrete with such sand, a large proportion of cementing material must be used to furnish the coating film, and the adhesion of the film to the smooth surface is comparatively weak. This sand should therefore only be used in default of better material.

106. Sea sand may be used for mortar making with either hydraulic lime or Portland cement, but the mortar should not be used in the erection of dwelling-houses, warehouses, &c., where the absorbent nature of the salt would tend to maintain a dampness in the walls. It is largely used in building dock walls and similar massive structures.

QUALITY OF SAND. 121

Sea sand is often deficient in the essential qualities of sharpness of angularity, and roughness of texture of surface of the grains, and sometimes is not entirely composed of grains of silica. If of suitable quality, and thoroughly washed in a stream of clean fresh water, sea sand can be used in mortar making for all purposes.

107. Loamy and clayey sands are injurious in cement mortar, and to a less degree in lime mortar. To use cement to the best advantage, the sand should be carefully selected, should have the grains angular, with rough texture, perfectly clean, and be not liable to decompose.

108. Crushed sandstone, vitreous blast furnace slag, or any vitrified rock such as vitreous lava, quartzite, or quartz rock itself, will furnish a good sand for mortar making. For a dense impervious mortar, it is best to use the finer particles as well as the coarse, but with an additional quantity of lime or cement above that which would be mixed with coarse sand only.

109. It appears that caustic lime and cement has no appreciable action on grains of silica, a mortar several years old has been examined under the microscope, and the angles of the quartz grains were observed to be intact. Conclusive proof of the absence of any action is, however, at present lacking; it is considered that there is always some free lime in mortar, and that this free lime may act upon the grains of sand.

110. Finely granulated blast furnace slag, and crushed vitreous slag is sometimes used as sand: and puddling and reheating furnace cinder is crushed small,

and used for mixing with fine concrete for face work. Scoriæ from furnace hearths and firebars are also used, and if of well vitrified durable quality, make a good sand.

Proportion of Sand used.

111. Gray chalk quicklime can be mixed with nearly three times its bulk of sand, an ordinary proportion is from 2 to $2\frac{1}{2}$ times the bulk of the lime before it is slaked.

Blue lias lime should not be mixed with more than twice its bulk of sand for a strong mortar, and the best mortar is made with equal proportions.

For rubble stone work at the Liverpool Docks, using Halkin hydraulic lime, the proportions were 1 of slaked lime to 2 of sand, and $\frac{1}{3}$ of crushed smithy ashes. For brickwork and for facework of thick walls 1 of slaked lime to 1 sand and 1 of smithy ashes.

Water used for Mixing.

112. The amount of water used for mixing good lime mortar should not exceed the quantity required to make a somewhat stiff and cohesive paste. If too little water be added the mortar will not set properly, and if too much water be used the mortar will be thin and sloppy, the sand is liable to settle and separate from the lime, cavities may be left in the mortar joints by shrinkage during setting, and it is said that the thin mortar never sets quite so hard as a thicker paste. Each kind of lime takes its proper

quantity of water, which can only be ascertained by practical trial, and only general rules can be given. The quality of the sand, and the size of the grains to be wetted, will influence the proportion of water. Due allowance must always be made for the porosity and absorptive nature of the bricks and stones to be cemented together, the mortar paste must not be robbed of any of the water required for setting; and allowance is also to be made for loss by evaporation in hot weather. It is a good plan to saturate with water all bricks and stones used in building, except in time of severe frost.

Common chalk lime may require about 6 gallons of water per bushel for mixing; a blue lias lime in the lump, from $2\frac{1}{2}$ to $4\frac{1}{2}$ gallons after slaking, and up to 6 or $7\frac{1}{2}$ gallons if the lime be ground, the water being always sprinkled in a state of fine division.

113. Thin liquid mortar is sometimes used in building the middle portion of massive masonry. The outer portions, or faces of the structure, are built for one or two layers, or courses, in advance of and above the hearting; the thickness of the facing varying between 9 and 14 inches of close jointed masonry in stiff mortar paste. The hearting is built with bricks or stones laid dry, i. e. open-jointed, liquid mortar, called grout, being then poured out of buckets and cans over the bricks and into the open joints. This is inferior work, even when the joints are carefully filled.

A better method is to fill the shallow basin enclosed by the facework with comparatively stiff mortar, softened if need be with a little water, and remixed with a long toothed rake called a "larry."

The bricks or stones are then well bedded by hand in the pool of mortar paste. This process is called "larrying"; that first described, "grouting."

114. A mass of pure lime mortar, if left unused for some time, may again be chafed (or cut and beaten up) with trowel or shovel, until it assumes its pasty plastic condition, and can then be used without serious detriment. This, however, is not the case with hydraulic limes, they should be used as soon as practicable after being thoroughly mixed, and if the mortar has set, it must not be again worked up for use; it can, however, be used, in hardened lumps, for foundation concrete work.

BLUE LIAS LIME MORTAR.

115. Blue lias lime may be purchased in the lump, or ground ready for use. Unless the user of the lime can place absolute reliance on the supply of the best quality only of ground lime, it is better to purchase the calcined lumps, to submit them to a searching inspection for underburned lumps, and to grind the lime at the works.

The lias lime should be carefully slaked by being wetted thoroughly on the clean floor of a shed, the layer of lime lumps being about 6 to 12 inches thick. A layer of clean sand, and straw mattrasses, &c., may be placed over the lime to retain the heat developed during slaking, the sand not to exceed the proportion of equal bulk for the strongest mortar. Not less than 7 nor more than 14 days may be allowed for slaking the lumps, and not less than 24 hours for the ground lime powder. The slaked lime must be sifted, unless it is to be ground in a mortar-mill.

After the complete slaking of the lime it may then be used for mixing into mortar. If mixed in the ordinary grinding mortar-mill, the sand is to be added at the last stage and to be crushed as little as possible, unless it is too large in the grain and worn smooth on the surface. In hand mixing the ingredients are to be thoroughly turned over on a clean platform, or hard floor, at least six times, and sufficient water to be used to furnish a pulpy mortar. The paste to be used soon after mixing, and no more mortar to be made at any time than can be used within the next 3 or 4 hours. Lias lime mortar that has begun to set and indurate must not be re-made for use, nor be mixed with new mortar.

The proportion of sand and other bodies, such as furnace ashes, or semi-vitrified burnt clay, used in the production of strong mortar, does not generally exceed 3 to 1 of lias lime, is often $2\frac{1}{2}$ and 2 to 1, and in especial cases for facework exposed to water $1\frac{1}{2}$ and 1 to 1. As an instance may be cited Aberthaw lias lime mortar used at Avonmouth Docks — 1 of lime to 2 of sand and furnace ashes equally mixed; the ingredients were ground in a mortar mill for 20 minutes to make mortar for rubble stonework, and for 40 minutes for brickwork.

Mortar Mixing by Hand.

116. For small quantities of mortar, the mixing is usually carried on by hand labour, spades or shovels

and rakes with long prongs being used. The sand and slaked lime should be well mixed in the dry state, being turned over with spadework three or four times on a clean practically impervious surface, generally a platform of boards. Then water is sprinkled on while the mixing continues, until the whole is a soft plastic uniformly mixed paste. About 6 or 8 turnings over are the minimum required. Ground blue lias lime is sometimes mixed with sand and water without previous slaking, then left for 6 or 8 hours, or longer, and then remixed, to ensure the complete slaking and wetting of all particles of lime.

Mortar Mixing by Machine.

117. When large quantities of mortar are required, mixing machines are generally employed. One form of machine in common use consists of a cast-iron circular pan or ring trough, which can be rotated horizontally round a central pillar. On opposite sides of the bed of the pan rest two cast-iron discs or rollers standing on edge; they rotate in a vertical plane round a horizontal common axis which has a limited extent of free vertical movement on the central standard round which the pan rotates.

Driving power is conducted to the pan by a fixed pinion wheel gearing into a circular or ring rack bolted to the underside of the pan; the discs (or edge runners) rotate by friction on the bed of the pan, or on such materials as are placed in the pan to be crushed.

The pan should be placed as low as possible and should be mounted in a substantial cast or wrought-iron frame on a firm foundation. The bearing of the pan at the foot of the vertical standard should be made in gun-metal or in hardened steel. In some cases the lower end of the vertical spindle is flanged out, and rests on a flat bearing surface, which is boxed in to contain the lubricant. Three parts of good machine oil to one of plumbago make a good lubricant.

Spiral adjustable springs are generally fitted in the framing on each end of the axle carrying the edge runners, to take off sudden shocks due to the runners having to surmount unusually large lumps of material. If the pan exceeds 8 feet in diameter, it is advisable to furnish additional support at the periphery by placing friction rollers beneath the pan. The edge runners and rollers should be fitted with removable tyres of hard metal, and the bottom of the pan should be also of hard metal and be easily removed and replaced. A false bottom is used, cast of specially durable metal, and fitted to the interior of the pan. The shaft carrying the pinion gearing into the ring rack may preferably be driven by a belt, on a pulley and on the fly-wheel of the engine.

Sometimes the runners or rollers are driven, and the pan is stationary. The vertical central shaft, and the horizontal axle carrying the rollers are rigidly connected, and a toothed wheel surmounts the top of the vertical shaft; this wheel, called a crown wheel, must be large, not less than 4 feet diameter, and must be securely keyed on to the shaft. It is driven by a small pinion wheel, and pulley and band. Pan driving is considered to be the best method.

118. The process of mixing in the edge runner mortar mills is to place the slaked lime in the pan and grind it under the runners for at least 3 minutes in a dry state, then water is added gradually, through a rose jet or a finely perforated tube, and finally the sand is put into the pan. The whole time of grinding is generally from 20 to 30 minutes, water being added slowly. If the lime be hydraulic, it is sometimes ground before being put into the mixing mill; the ground lime must be exposed to air slaking for several days, generally at least seven, it is placed in a thin layer on the floor of a dry shed. Hydraulic lime in the lump must be allowed to slake for a period of at least 2 to 7 days before mixing.

A mistake commonly made in mixing mortar in an edge runner mill is that of adding the sand at an early stage of the process. If the sand has been selected of the right size of grain, angularity, and roughness of texture, it is a mistake to crush that sand into finer particles under the heavy edge runners. The smaller the grains of sand, the larger will be the total area of surfaces in a given bulk, and the more lime will be required to form a film enveloping each grain.

For the lime, on the other hand, the finer it is ground the more perfect will be the enveloping film of lime on the surface of each sand grain, and the greater the area of surface that a definite quantity of lime will cover.

Hence the sand should be added at the final stage of grinding, which should be continued no longer than is absolutely necessary for perfect mixing alone. If the sand be mixed with the ground lime in a machine such as a pug-mill, which mixes and does not crush, the mortar will be properly prepared.

119. If pozzolana, semi-vitrified brick, or similar burnt clay or forge ashes be added to the mortar, it should be ground for as long a time as the lime, that is, put in the mill pan at the same time as the lime. A good plan is to grind in the mill the lime, &c., and to mix by hand the sand with the ground lime pulp, or to use a mechanical mixer for the separate amalgamation of the lime and sand.

The proportion of pozzolana, or similar volcanic ash, added to rich or to lias limes varies with the quality of the work. To a rich lime $2\frac{1}{2}$ times its bulk of pozzolana may be added for a strong mortar, ranging thence down to equal bulks of lime, pozzolana, and sand. To a lias lime, one to one and a half times its bulk may be added for use in sea-water, but the precise quantities should be fixed after analysis of the ingredients and after experimental trials.

120. French engineers sometimes employ a mixing machine, consisting of a long hollow cylinder, or semi-cylinder, in the axis of which is mounted, to rotate, a shaft round which is a continuous broad blade in a spiral, as in the Archimedean screw. The mixer is fixed at a slight inclination to the horizontal; the rotation of the shaft and blade mixes the ingredients placed in the cylinder, and the mixed mortar "creeps" to the lower end of the machine, where it drops into a barrow, tram-wagon, &c.

Strength of Lime Mortars.

121. Burnell gives for strength of ordinary mortar,

(1) Resistance to tensile stress .. .. 14 lbs. per sq. inch.
(2) Resistance to crushing stress .. 42 lbs. ,, ,, ,,
(3) Resistance to stress causing sliding 5¼ lbs. ,, ,, ,,

It is exceedingly difficult to state any general measure of adhesive strength of mortar, as so much depends upon the quality of the lime, the quality of the sand, the nature of the material to be cemented together, and the character of its surface. Mr. Mann states (Proc. Inst. C.E., vol. lxxi.) the adhesion of Portland cement to the surfaces of the various materials mentioned, the values being in lbs. per square inch of surface.

Adhesion to slate .. .. at 7 days, about 50 ; 28 days, 80 lbs.
,, { Portland limestone } ,, ,, 26 ; ,, 60 ,,
,, { Granite, chiselled surface } ,, ,, 41 ; ,, 97 ,,
,, { Limestone, crystalline } ,, ,, 57 ; ,, 93 ,,

Cement Mortar.

122. For masonry structures exposed to the action of running water, or of waves, it is necessary to use a quick-setting mortar not liable to suffer deterioration from constant submersion, or from alternate wetness and dryness. Portland cement is generally used in such cases, but the workman finds some difficulty in

using mortar made of Portland cement and sharp sand only; it does not spread freely, "works short" as the man terms its defect. To remedy this inconvenience, sometimes loam (clayey sand) is added, or even a little earth or vegetable mould; the mortar is then more easily spread, but the added matter weakens it materially, and the practice should be prohibited.

A comparatively harmless remedy is to add a little whiting, or a good chalk lime, to the cement, in the proportion of about $\frac{1}{12}$ the bulk of the sand used. Another proportion is to add to 1 of cement $\frac{1}{4}$ of dry slaked lime, this cures shortness; any added lime must be thoroughly slaked.

A better method is to add blue lias lime, or even a good gray chalk (stone) lime. "Short" mortar may also be cured by adding one-sixth of its bulk of crushed well-burned brick, partially vitrified, and not sandy. Partially vitrified burnt clay (burnt ballast) of good quality can also be used. Shortness of the mortar can also be cured by long continued mixing and beating with a shovel.

A mortar composed of 1 of Portland cement, $\frac{2}{3}$ of stone lime and 8 of sand is said to be superior to ordinary gray chalk lime mortar composed of 2 sand to 1 of lime.

123. A loamy sand gives a freely spreading mortar, but as there is a considerable loss of strength when any clayey matter is mixed with cement, the use of loamy sand is detrimental. It has been used in certain cases with the avowed object of making an easy working mortar, and giving to it greater consistency and density. As an example, a mortar for masonry

dock walls was made of 1 of Portland cement to 7 of clean coarse sand, and 1 of foundry sand (loamy fine grained sand) containing about 10 per cent. of clayey matter. It was noticed that fine grained sand made a poor mortar, and that sand whose grains were coated with loam did not make so good a mortar as coarse clean sand, with the fine loamy sand added for the purpose of producing an easy working mortar. The mortar would have been stronger had the loamy sand been omitted, and gray chalk or blue lias lime substituted.

A good cement mortar is made of $2\frac{1}{2}$ clean sharp sand to 1 measure of Portland cement; or, 2 sand, $\frac{1}{2}$ of ground clinker or vitreous slag to 1 of cement. Sugar, or molasses syrup, is added to Portland cement mortar in hot countries; the coarser the sugar syrup or the molasses the better appears to be the result; about 1 lb. to 1 bushel of cement. Molasses retards the drying and setting of cement more than fine sugar. Another proportion is given as from $\frac{1}{8}$ to 1 per cent. by weight of molasses, according to the results of experiments; the strength of the mortar is said to be increased by 15 to 20 per cent. in 3 to 4 months.

124. For a strong mortar, the proportion of 4 of sand to 1 of cement should not be exceeded, and for a higher class, the maximum of 3 to 1 may be adopted. Water is to be used in the proportion of the sand, and with regard to its quality; due allowance must always be made for the particular usage of the mortar, for the porosity of the materials to be cemented together, and for the loss of moisture by evaporation and other causes. Mortar must not be

CEMENT MORTAR.

allowed to dry by evaporation of the contained moisture. One of sand to one of cement by bulk will need about three gallons of water to the bushel of cement; two of sand to one of cement, about five gallons; and three of sand to one of cement about six gallons. Sufficient water should be used to thoroughly wet the sand, &c., and to saturate the cement, stopping short of making a sloppy mortar.

All mortars made with highly hydraulic limes and cements must be mixed in such quantity only as can be used within one to three or four hours after mixing; some cements must be used immediately after mixing. The materials should be well mixed in the dry state, and then water should be sprinkled on through a finely perforated rose jet.

125. For use in sea water in an exposed position a cement mortar of 1 cement to $1\frac{1}{2}$ sand is good and practically watertight, and as a maximum of cement for a dense impermeable mortar, the proportions of 1 to 1. Michaelis recommends 2 of fine sand to 1 cement, Mr. Sandeman a ratio of $1\frac{1}{2}$ sand to 1 cement, as giving an impermeable mortar. A mortar of 1 cement to 3 of sand was found to be injured by percolation of sea-water.

At the Vyrnwy Reservoir masonry dam, the cement mortar in the foundations was 1 cement to 2 sand; for the upper part 1 to 3; and for the road viaduct on the top, 1 to $3\frac{1}{2}$.

Cement mortar made with an excess of coarse sand may be permeable to water. By adding lias or stone lime to the cement, the permeability may be greatly diminished. A mixture of 1 cement to 2 slaked lime to 6 of sand, was found to make an impermeable slab. Another proportion is given as 1 cement to 2½ sand and ¼ of slaked lime for a stronger mortar than the foregoing.

For cement mortar in building an arch of masonry the proportions of 1 cement to 3 sand were adopted.

For strength the mortar joints should never be more than ¼-inch thick, and there seems to be no advantage in this respect in mortar joints of less than ⅛-inch thick.

Mortar-mixing in Time of Frost.

126. It is a general custom to cease using mortar or concrete when exposed to a temperature at which the water for hydration of the lime or cement is liable to become frozen quickly; the lime will be deprived of some of the water by its becoming frozen, and the expansion of the frozen water is liable to cause disintegration of the mortar or concrete. Severe frost is therefore detrimental to the mixing and use of mortar and concrete. Salt is sometimes added to prevent freezing, being dissolved in the water; but this plan should not be adopted for warehouses or dwelling-houses.

Building may however be carried on without any great risk during severe frost, if the stones or bricks be exposed to a jet of steam for from 15 to 20 minutes, and if boiling water be used for mixing the mortar. Old surfaces must be well steamed before new masonry is added, and new work must be covered over and protected from wind by screens, &c., as soon as finished, and be well covered up at night. Concrete has been thus laid down with the thermometer at 22° F., and no serious injury ensued except to about 2 inches depth at an exposed surface.

A method said to be adopted at Christiania (Norway) for building during severe frost, down to 14° F., consists in the use of unslaked lime. The mortar is made in small quantities only, from unslaked lime, and used at once, and the greater the cold, the larger the proportion of lime in the mortar. The bricks used must be dry, and the mortar is probably used in a very moist condition and in thin joints. The new work is always protected as soon as built, especially against rain, snow, and cold winds.

From some experiments cited in Proc. Inst. C.E., vol. c. p. 425, it appears that the addition of a solution of crystallised soda has the effect of enabling a cement or lime mortar to resist the effects of very severe frost. The solution used was 1 kilo. (2·2 lbs.) of soda-crystals to 2 litres (0·44 gallon), or to 3 litres of water. Another proportion is 1 lb. of anhydrous soda-crystals to a gallon of water. This experiment needs additional investigation.

Plastering.

127. Strong cement mortar is used as a coating to the surface of walls, or of a mass of concrete of inferior strength and density. In coating masonry walls, the surface of the wall should be washed clean, it sometimes needs to be scrubbed with wire brushes; it should be damp when the coating is applied, the

bricks or stones having been well drenched with water thrown on ; and the adhesion of the coating is assisted by the dusting on to the wetted surface of a little dry cement powder.

When used as a vertical facing to a mass of inferior quality, the facing mortar (or fine concrete) and the inferior backing material should be deposited at the same time, being temporarily separated by a thin board, or an iron sheet. The barrier being promptly taken out, the two materials are at once incorporated by chopping with a spade on and across the junction line of the facing mortar and the backing.

Only the best cement and the best quality of sand, or small gravel, &c., should be used for the facing stuff, and the mixing must be thoroughly performed. Crushed glass, vitreous slag, quartz rock, quartzite, ancient lava and similar igneous rocks are sometimes used. Special care must be taken to ensure cleanliness of surfaces and of materials.

CEMENT FACING ON BRICKWORK. POINTING OF JOINTS.

128. The joints should be raked out clean for 1 inch in depth, then soaked with water by a hydropult or garden engine. One part cement to three of sand, mixed thoroughly, should then be laid on and pressed well into the joints. The edges of the cement laid on must be kept wet, and when the facing has set firmly it should be wetted again, so as to indurate slowly.

Cement and Lime Mixed Mortar.

129. In building massive works it is sometimes desirable to use a strong and quick setting mortar for the facework only. Such mortar can be made by adding Portland cement to the lime in general use. In one case, one measure by bulk of Portland cement was added to one of Burham gray chalk lime, and five to six measures of good sand. The cement should be well mixed with the lime in dry powder.

Another instance gives 1 cement to 3 sand and $\frac{1}{2}$ of a good slaked lime, this proportion gives a good dense mortar for foundation and above-ground work.

Strength of Cement Mortar.

130. The addition of sand to Portland cement mortar lessens its cohesive strength in proportion to the amount of sand. The diminution in strength may be from 20 to 10 per cent. for each proportion of sand, according to tests made on briquettes seven months old. In another case, a mortar of one sand to one cement gave in two years' time an ultimate strength of only two-thirds of the neat cement.

The strength of a cement mortar, as usually tested by tensile stress, largely depends upon the quality of the sand, its purity, size, form, and texture of surface of the grains; hence no definite ratio of deterioration can be accurately stated for ordinary mortar mixing.

Mortar of one cement to two of sand has about one-half the strength of neat cement mortar.

Adhesion of Cement Mortar.

131. As a test of the adhesive strength of cement mortar, some good ordinary bricks were cemented together with a mortar of 2 sand to 1 of Portland cement. The mortar was allowed to harden for 28 days, the cemented surfaces had an area of about 40 square inches. The adhesion ranged from 15 to 30 lbs. on the square inch, according to the character and texture of the surface of the bricks.

In the case of glazed surfaces, such as the highly vitrified surface of some of the blue Staffordshire bricks, or glazed tiles, it is found that Portland cement has comparatively feeble adhesion.

CHAPTER VII.

CONCRETE; PROPORTION OF INGREDIENTS; HAND AND MACHINE MIXING; LIME CONCRETE.

CONCRETE.

132. CONCRETE is a mixture of either hydraulic lime or Portland or Roman cement and gravel (consisting of siliceous sand and more or less water-worn stones, and generally some loamy sand), or other material such as broken stone, broken brick, &c. ; in fact any angular or even rounded fragments and pebbles of durable nature whose surfaces are sufficiently rough and granular to furnish points of adhesion for the indurating particles of cement.

In a concrete structure, such as a quay, dock, or retaining wall, required to be impervious to percolation of water, the vacant spaces between the angular or rounded fragments of which the concrete is composed must be densely filled. The best material for this purpose is clean mixed-grain sand, of sizes varying from $\frac{1}{24}$ or $\frac{1}{16}$ to an $\frac{1}{8}$ of an inch diameter. A fine sand is not so good as a mixture of coarse and fine grains, the latter not exceeding one-third of the bulk.

Sufficient cement should be used to furnish a coating to the surface of each particle of the matrix, and the extent of surface of the particles as compared with their bulk is much greater with fine than with coarse sand. For instance, a 3-inch sphere has one-eighth the bulk of a 6-inch sphere, but its surface-area is one-fourth that of the 6-inch.

To ensure the perfect coating of each particle, a larger proportion of cement must be taken with fine than with coarse sand. If the proportion of cement taken for fine sand be that sufficient only for coarse sand, the cement films enveloping the particles will probably be imperfect, and the concrete will be deficient in strength. Fine sand concrete, therefore, requires more cement, and probably costs more, than coarse sand concrete of equal strength.

A method of ascertaining the probable quantity of sand required to make an impervious concrete is to fill a large water-tight tank (of say a cubic yard capacity) with impervious pebbles or rock fragments only, and then add water in measured quantities until it rises to the rim of the box; the cubic measure of water will give, approximately, the measure of the spaces to be filled by the sand. About 5 per cent. should be added to allow for the shrinking of dry sand when wetted; and another allowance may have to be made for permeability of the rock fragments.

133. In many cases a comparatively open textured, or honeycombed, concrete may be as serviceable as a dense concrete, and will be more cheaply made, the cost of the cement being generally the more important item. A concrete composed mainly of angular or rounded fragments will contain a relatively

small area of surfaces to be coated, and a minimum quantity of cement can be used. If sand were added to this concrete made with a minimum of cement, the result would be to weaken the concrete to an extent proportional to the amount of sand used, and to the fineness of its grains. An open textured concrete of 1 of cement to 10 of stones is serviceable.

A short series of experiments made by Mr. Darnton Hutton during the construction of the concrete walls of the North Sea Canal Harbour at Ymuiden, illustrates clearly the weakening effect of the presence of fine sand in a concrete poor in cement. (Proc. Inst. C.E., vol. lxii. p. 196.)

134. In specifying the proportions to be used in mixing concrete, the sand should always be distinctly stated apart from the stones or pebbles. For instance, a concrete was specified to be made of 12 of shingle to 1 of cement. This shingle was found to contain in some cases nearly one-third its bulk of sand, and the concrete made in the stated proportions would be either serviceable, or dangerously poor in cement, according to the small or large amount of sand in the shingle; if the sand were plentiful the concrete would be 16 to 1 instead of 12 to 1. The concrete should have been specified as 1 cement to 3 of sand and 9 of clean stones, making a proportion of 1 to 12.

135. Any compact crystalline stones may be used for concrete that is not to be exposed to severe compressive or other stresses, nor to strong disintegrating or decomposing influences. The most suitable broken stone has a granular surface, is coarse-grained,

compact and highly siliceous, and the texture of the surface of the grains must be rough. For use in sea-water it must be practically impervious, and not liable to be acted upon by the sea-water, nor by exposure to the weather, nor to acid-laden air, nor be burrowed into by boring molluscs. In some instances concrete must be fire-resisting. Granite, syenite, and some of the more siliceous igneous rocks, quartzites, and highly siliceous sandstones are well suited for important work. Limestones are liable to decompose and also to be attacked by boring molluscs; sandstones with calcareous cement are liable to disintegrate; slates and shales, unless indurated to the point of incipient vitrifaction, are liable to decompose; these and similar stones must only be used in works where their weakness will not be attacked, or will not be detrimental. Sandstones with siliceous cement are the best in resisting the joint action of fire and water.

135A. The stones are broken into fragments, either by hand with hammers, or, where large quantities are required, by a machine driven by a steam or gas engine, or by water power. The best form of stone-breaker consists of two very strong slabs of iron or steel having vertical corrugations on the breaking face; one slab is fixed, the other is movable and is made to approach and recede from the fixed slab to a definite extent, so that stones can be broken to any desired size. Small chips and undersized fragments are separated by passing the stones as they come from the breaker through the interior of a hollow cylindrical sieve 6 feet or more in length, the shell being pierced with a series of graduated holes. A hard granular sandstone can be broken in a stone-breaker to yield stone for concrete, and the smaller pieces can be crushed under rollers to furnish a sand well adapted for either mortar or concrete mixing.

136. River ballast and pit ballast often contain more or less loamy matter, which should be carefully washed away for good concrete; a quantity not exceeding $1\frac{1}{2}$ per cent. will not be injurious for ordinary work. An excess of loam not only weakens the cement, but also renders the concrete liable to injury from frost, probably owing to the moisture absorbed and retained by the clayey matter.

137. Broken bricks, tiles, unglazed earthenware, &c., may be used, provided they are well burned, and not vitrified to a smooth glassy surface; they are well adapted for fire-resisting concrete. Soft sandy bricks, and bricks of light weight, may be useful for making so-called fire-proof floors, but the concrete will not be strong to resist compression. Burnt clay, as generally produced, is insufficiently burned to make a durable concrete; the burning should be carried to incipient vitrifaction, and the colour of the product will generally range between dull dark red and dull purplish red.

138. For a very light concrete, for flooring, &c., crushed coke is sometimes used, also crushed boiler furnace cinder, and blast furnace slag that has been granulated by running it from the furnace into water. Concrete made of such porous materials will generally permit nails to be driven in, so that floor boards, or

floor joists, can be nailed down on the top of the concrete; its strength to resist compression is only moderate. Coke concrete is a good conductor of heat, and, containing carbonaceous matter, should not be used in contact with wooden fittings in places where the concrete may be accidentally exposed to fire, as under hearths, &c.

Coal cinders and lime concrete (1 lime to 4 cinders) has been largely used for all kinds of construction. The concrete is light, yet strong, if good lime or cement be used; its weight is about 2080 lbs. per cubic yard, and it has been used for the following arches: 21 feet span, 4 feet rise, thickness at crown and at haunch, $1\frac{1}{2}$ feet and 3 feet respectively; the load carried was 2 tons per square yard distributed, or nearly 500 lbs. per square foot. Another arch, span 16 feet, thickness at crown 14 to 16 inches. The material was well rammed into the moulds. (Proc. Inst. C.E., vol. lxxxi. p. 352.)

Granulated slag is prepared by directing the flow of molten slag into a wheel, or drum, about 14 feet diameter, rotating in a vertical plane, with the lower part of the rim running in water, of a depth of 18 to 24 inches. The arms of the drum are curved from the rim to the boss, so that the spout for the molten slag, and the slag sand shoot, can deliver into, and receive from, the concavity of the rim. A full description of methods of utilising slag is to be found in papers by Mr. Charles Wood in the 'Minutes of the Proceedings of the Iron and Steel Institute,' for 1873 and 1877; and also in the 'Society of Arts Journal' for May 14th, 1880, and by Mr. G. Redgrave in the same journal for January 31st, 1890; also in a paper by Mr. P. L. Simmonds, in the same journal for December 22nd, 1882.

Mixing of Concrete.

139. Portland cement concrete is prepared by mixing with the unit quantity of cement certain measured proportions of the other ingredients; the mixing is first carried on with dry materials, and then during the gradual and regular addition of a measured quantity of water. The mixing is carried on either by hand labour, or with the aid of mixing machines driven by hand or steam power.

Hand Mixing.

140. The ingredients should be mixed on a platform (of boards, &c.) which can be kept free from dirt. A movable platform may be made, in two portions, of boards nailed down to two or three parallel ledger timbers placed beneath. When the two portions are placed side by side the platform should be not less than 9 feet, or larger than 12 feet, square. The division of the platform into two parts facilitates its handling and removal.

Each of the ledgers to which the boards of each half are nailed should project on one side at least six inches, and the projecting ledgers of one half platform should just slide past the corresponding ledgers of the other half, when the two portions are placed together to make the mixing platform. The boards

L

should be $1\frac{1}{4}$ or $1\frac{1}{2}$ inches thick, and the ledgers 5 by $1\frac{1}{2}$ inches, the nails are punched down about half-way through the boards, and clenched on the under side of the ledgers.

141. The gravel or broken stone is usually measured in a wooden frame or box, without bottom or cover, which is placed on the platform and filled with the gravel or stone. The box may have internal dimensions giving a capacity of a cubic yard, or of some fraction. As Portland cement is most conveniently transported in sacks or bags containing two bushels each, or in barrels holding about 375 lbs., the capacity of the box measure for the gravel, &c., may conveniently be made to meet the fixed proportions of the ingredients, a 2-bushel bagful of cement being added to one boxful of gravel, &c. The bushel contains 1·283 cubic feet, and the weight of the cement will be from 90 to 100 or 110 lbs.

Another plan is to make the box of the capacity of, say, half a cubic yard, and having a depth from top to bottom of twice as many inches as the sum of the proportions of cement to gravel. Thus, for seven to one concrete, the box would be made 16 inches deep. At 2 inches below the top of the box, a lath is nailed on all round inside; the gravel is filled into the box up to the level of the top of the lath, and the remaining 2 inches with cement. The box is now lifted up and taken away for re-use, the ingredients fall into the shape of a truncated pyramid, and are at once turned over by at least two men with spades, working on opposite sides of the heap, and at the same time are raked about with a long two-tined rake.

HAND MIXING.

The men turn over the materials and shovel them up into a conical heap. This turning over completely, and raking about the materials, is carried on at least twice in the dry state; the prescribed quantity of water is then added in a finely divided state, through a rose-jet on a hose or watering-pot; no more water must be added than will just form a pulpy mass— about 11 to 12 per cent. by bulk is generally sufficient. During the gradual addition of the water, the mixing is carried on till the materials have been completely turned over, in all directions, at least five times. They should be then sufficiently well mixed for instant use in constructing the work. If there be any liability to loss of water by soakage, &c., a superabundance should be used, it will quickly drain away from the concrete if not required for hydration.

The strongest concrete is made by mixing cement and sand to a pulpy paste, and then thoroughly incorporating with the mortar the clean pebbles or stones.

In a gang of mixers for producing a large quantity of concrete, the following distribution may be made.

Getting stone and sand	3 men
Getting cement and measuring	2 ,,
Gauge box fillers	2 ,,
Mixers	6 ,,
Boy to handle water hose	1 boy
	13 men, 1 boy.

Such a gang of men can turn out 35 cubic yards of concrete in 10 hours. The mixers will be in two parties of three each, two with shovels, and one raking.

PROPORTION OF INGREDIENTS FOR CEMENT CONCRETE.

142. The proportion of ingredients of cement concrete depends upon the class of work to be executed, and upon the quality and fineness of grinding of the cement powder.

For concrete in foundations fairly dry, a proportion by bulk of 1 cement to 6 clean gravel pebbles and 2 of sand, or 1 cement to 8 to 10 parts of broken stone; another proportion is 7 shingle to 2 sand to 1 cement for ordinary concrete; for foundations on which may come heavy shearing forces, 1 of cement to 4 gravel pebbles and 2 sand, or 1 cement to 5 or 6 of broken stone. A good concrete has been made of 1 cement to 2 of crushed quartz rock sand and 4 of quartz fragments up to 2-inch size.

For a reservoir wall, to be watertight, the proportions may be 1 of cement to 4 gravel pebbles and 2 of sand if the cement be very finely ground. Another proportion is 1 cement to $5\frac{1}{2}$ broken stone and $1\frac{3}{4}$ of sand, but this is a maximum of materials. For the ordinary cement in the market, 1 cement to $3\frac{1}{2}$ gravel pebbles and $1\frac{3}{4}$ sand ; or 1 cement to $3\frac{1}{2}$ broken stone and $1\frac{1}{2}$ sand. Another authority fixes the maximum of materials for ordinary cement at 1 cement to 3 stone and $1\frac{1}{2}$ sand, and for coarsely ground cement 1 cement to $2\frac{1}{2}$ stones and $1\frac{1}{2}$ sand.

The sand and cement when mixed should be about one-third of the total bulk of the concrete, and about one-half of the bulk of the stones, which are to be of varying sizes. The sand to be not more than twice the bulk of the cement.

For concrete blocks exposed to sea-waves (facing blocks), 1 cement to 5 clean shingle and 2 of sand; for hearting blocks not exposed to the sea, 1 cement to 10 or 12 clean shingle, or 1 cement to 3 sand and 9 pebbles for backing of dock walls. A better concrete would be given by 1 cement to 2 sand and 8 of pebbles.

In another instance, for concrete blocks not exposed to abrasion, 1 cement to $2\frac{1}{2}$ sand and $6\frac{1}{2}$ clean gravel pebbles or shingle. These proportions require $2\frac{1}{2}$ bushels of cement to a cubic yard of mixed concrete. A cheaper concrete for foundations, or for backing to stronger face work, may be made of 2 bushels of cement to the cubic yard of concrete; and to make a stronger concrete for facework to resist abrasion, 3 bushels of cement; the proportion of cement in the last concrete will be 1 in 7, nearly.

Another proportion for ordinary concrete for foundation work is $4\frac{1}{2}$ cubic feet of cement to 25 to 27 cubic feet of broken stone or pebbles, and 9 cubic feet of sand with 25 gallons of water for mixing, make 1 cubic yard of concrete. Or 1 barrel of Portland cement (375 lbs. net), 5 barrels of broken stone, &c., and 2 barrels of sand will make about 20 cubic feet of concrete.

From 22 to 25 gallons of water are used in mixing a cubic yard of ordinary concrete, that is about 1 part by volume of water to $6\frac{3}{4}$ or 8 of concrete. But the quantity of water used depends upon the absorptive nature of the materials, the proportion is given for ordinary impermeable gravel pebbles. The water must be perfectly clean (free from all muddiness) and should not contain acids, especially sulphuric acid.

Sometimes the quantity of water is specified at not less than 40 lbs. weight, or 4 gallons to the cubic foot of cement, which may weigh from 75 to 95 lbs., according to its fineness, coarseness, &c., but it is better to take into consideration that the surfaces of the stones and of the grains of sand must be wetted. Mr. Draper states that 41 per cent. by weight of water is required for the complete hydration of cement, according to its theoretical composition. Another proportion of water is 18 gallons to the cubic yard of materials, taking the sum of their respective volumes.

Using a full quantity of water to thoroughly saturate the cement and wet the materials, it is found that about one-eighth more sand and shingle can be got into the measuring box than when the materials are put in dry. If allowance be made for the shrinkage of the wetted materials a full volume of mixed concrete will be obtained. On the other hand, if no allowance of additional materials be made to the measured volume in the dry state, the mixed concrete will be deficient in volume by about 10 to 12 per cent. In dry-mixed concrete there will be larger air spaces than in well wetted concrete.

143. Abstraction of moisture by external heat, or wind draughts, or by capillarity from properly mixed concrete is most prejudicial, and should be counteracted by keeping the concrete wetted on the surface while setting. Some engineers use an excess of water

in mixing, to allow for loss ; but the strongest concrete is made with the proper proportion of water, the mass being kept wet on its surface for three or four weeks after deposition.

144. In building thin concrete walls for houses, &c., timber or iron frames are erected *in sitû* to form moulds into which the soft concrete can be deposited and allowed to remain till the concrete has hardened. The frames generally used are subject to patent rights ; thin sheets of iron or planed boards are fixed in the framework to give a smooth surface to the concrete, and their surface is also coated with a thin wash of clay mud to prevent any adhesion of the concrete to the mould. Another preventive is a paint of common yellow soap, dissolved in water, boiled and stirred. For 9-inch walls the pebbles or stones should range up to $\frac{3}{4}$ and 1 inch diameter, and for thicker walls of similar character, up to 3 inches diameter for an 18-inch wall. No stones or lumps to be placed nearer than 3 inches from face of wall. The concrete must be well rammed down, and made as dense as possible in the mould. Provision must be made for the insertion in the mould of all woodwork (such as door and window-frames, wood-bricks, &c.), that is built in ordinary brickwork. (See Art. 184 and "Concrete, its use in Building," by T. Potter.)

145. A poor concrete is sometimes used where it is required merely to fill up pockets in a structure, or to give weight and mass. Such concrete may be made in the proportions of 1 cement to 10, 12, or even 15 of broken stones, but for these concretes the

cement must be finely ground, and there must be little or no sand; with coarse cement, and with sand in appreciable quantity, this proportion cannot be trusted to give a durable concrete. Concrete for pockets must be made of a cement that will not expand in setting, or rupture of the walls may be caused.

146. For the construction of a bridge in Switzerland, a concrete of 250 to 340 lbs. of cement to the cubic yard of concrete was used in the foundations, and 420 lbs. in the superstructure; or 1 to 14·8 and 1 to 10·9 by weight for foundations, and 1 to 8·8 for the superstructure. Taking a weight of 90 lbs. of cement to the cubic foot, the proportions are respectively 1 in 9·7 and 1 in 7·14, and for the superstructure 1 in 5·78 by bulk. The span of the bridge on the skew of 71° 40′ was 63 feet 4 inches; on the square, 59 feet 9 inches; the rise of the arch, 4 feet 7 inches; thickness of arch at crown 3 feet 3½ inches; at the haunches 3 feet 11 inches. The thickness of the abutment at the middle, 14 feet 9 inches, and at the ends under the parapet walls 26 feet 3 inches, measured along the outer face of the bridge. The invert arch was 1 foot 4 inches thick, and its versed sine 1 foot 8 inches. The concrete was made of lake gravel and the best local Portland cement. See Proc. Inst. C.E., vol. lviii. p. 311 and plate 7. Another example of concrete bridges is cited in Proc. Inst. C.E., vol. lxxix. p. 394.

147. For sewers and culverts, large conduits, &c., the proportion may be 1 cement to 6 of coarse clean sand from a river. For steps, paving, &c., 1 cement to

2 parts of fine gravel (to pass through ¼-inch mesh) to make a strong facing about 1 inch thick. The gravel to be of all sizes from under ¼ inch to $\frac{1}{16}$ inch, and to be clean and to contain but little sand.

A poor concrete, such as 1 cement to 12 of gravel pebbles, is often faced with from 9 inches to 24 or 30 inches in thickness of better concrete. As an instance, at Chatham Dockyard, Mr. Bernays used a facing of about 1 cement to 2 coarse sand and 4 of puddle or reheating furnace cinder, crushed to ¾ inch cubes; facing about 10 inches thick on some of the dock walls, the facing being thoroughly incorporated with the main mass of concrete.

In another instance a sea wall was faced with 6 inches of stronger concrete made of 1 cement to 4 of whinstone (old lava rock) broken to ½ inch cubes.

A watertight rendering or coating to ordinary concrete, or to brick or stone masonry, is sometimes made of 1 cement to 1 sand and ¼ to $\frac{1}{6}$ of good slaked lime; thoroughly sifted gray chalk is better than a pure chalk lime. The coating must be at least ¾ inch thick, and thicker according to the amount of exposure and of abrasion it may have to resist, and it must be thoroughly incorporated with the main body of concrete. This coating is to be compressed with trowel work, and made as dense as possible, the surface to be worked to a closeness of texture that resembles a glaze.

For a facing concrete to a riverside quay wall, 6 inches thickness of 1 cement to 4 of ballast has been used.

EFFECT OF AGE ON PORTLAND CEMENT CONCRETE.

148. It is probable that Portland cement concrete changes its nature somewhat with age, its hardness and brittleness increase, resistance to crushing increases while its tensile resistance lessens.

MACHINE MIXING OF CONCRETE.

149. Machines for mixing concrete are much used where large quantities of concrete must be made in a limited time. The machines may be stationary, or mounted on a wheeled framework running on a tramway so that the mixer can be brought close to the place where the concrete is deposited. Stationary machines are used in a block-making yard, also when the out-turn of concrete must be large; the mixed concrete is delivered direct from the machine into trucks which are taken along a tramway to the place of deposition. Movable machines are generally small and are worked by hand. There are many different varieties of mixers, but they may be classed into two main divisions. (1) A hollow cylinder (or semi-cylinder) either mounted vertical, or slightly inclined from the horizontal; in the axis of the hollow cylinder is mounted a shaft carrying arms for stirring and mixing, and the cylinder may be either fixed, or made to rotate in the same or in a contrary direction to that of the central shaft. (2) A closed chamber, mounted on trunnions or on a shaft, and

rotated ; the materials placed inside the chamber are so turned over, lifted and let fall, that comparatively few rotations of the chamber ensure perfect mixing. The shapes of the chamber differ greatly.

Stoney's Mixer.

150. This is in the form of a trough, from 7 to 9 feet long and $3\frac{1}{2}$ feet wide, of cast or wrought iron, and it is mounted on an iron or timber framing, being supported at each end by wrought or cast-iron brackets or knees bolted or riveted to the exterior of the trough and to the framework. The trough is semi-circular in section in the lower half, the upper sides being splayed out. In the axis of the lower half is a wrought-iron shaft, $3\frac{1}{2}$ to 4 inches diameter, or square, on which are mounted radial blades of wrought iron at regular intervals in a spiral round the shaft. The blades are slightly inclined so as to push forward as well as mix the concrete when the shaft is rotated. The inclination of the blades can be altered so as to accelerate or retard the delivery of the mixed concrete at the end of the trough, which is mounted with a slight fall towards the delivery end.

The gravel and cement in their proper proportions are shovelled into the upper end of the trough, or are delivered through a hopper-shoot fixed above the mixer, which is often placed in a two-story shed, the upper floor being a materials storage floor, with a hopper above the mixer. The first three or four blades of the rotating shaft turn over and incorporate the materials, which then pass the water supply rose-jet, placed at a little more than one-third of the length of the trough. Thence the moistened materials gradually travel to the delivery end of the trough; the mixing should be very complete, and the concrete coming from the end of the trough homogeneous in quality and uniform in colour.

An advantage of this simple machine is that comparatively large stones can pass freely between the blades; worn-out blades can easily be replaced, and the supply of water can be adjusted in quantity and in regularity of supply. The machine is described and illustrated in Proc. Inst. C.E., vol. xxxvii.

MESSRS. CAREY AND LATHAM'S MIXER.

151. This mixer is a hollow cylinder of wrought iron or of mild steel. In the axis of the cylinder is a shaft carrying curved arms and scrapers, this shaft can be rotated in the same direction as, but at double the speed of, the cylinder itself, which is fixed slightly inclined from the horizontal on a strong framework; sometimes the framework is mounted on wheels running on a tramway. The cylinder rests on friction rollers, in contact with strong bands (of rectangular bar) encircling the cylinder.

At the reception end of the cylinder are two pocket-wheels, or bucket chains or bands, one on each side, which are worked at the same time as the cylinder; the sizes of the buckets are proportioned to the gravel (or broken stone) and the sand, and the buckets scoop up sand or gravel from a heap and deliver their contents into a hopper mounted on each side at the reception end. Above this hopper is a raised platform on which the cement in bags is stored for use; a bag is emptied into a central hopper raised above the others.

The cement, gravel and sand fall into the hopper compartment leading to the cylindrical chamber, and are forced along by a broad spiral blade round the shaft till the mixing arms are reached. Water is supplied about mid-length of the cylinder. This machine automatically measures the gravel and sand; one pocket-wheel has an adjustable side, so that a certain range of variation of quantity is possible.

This is an excellent mixer, it can turn out a large quantity of mixed concrete in a short time, and is extensively used. It is fully described and illustrated in 'Engineering,' July 3rd, 1885, in an article on the Newhaven Harbour Works. Also in Proc. Inst. C.E., vol. lxxxvii.

Messrs. Lee & Co.'s Concrete-Mixer.

152. This mixer consists of a cast-iron hollow cylindrical case, mounted diagonally on a 4½-inch wrought-iron horizontal shaft; the chamber or case may be about 6 feet long and have a diameter of 3½ feet, the thickness of the metal being about ¾ inch. The axis of the hollow cylinder is inclined at an angle of about 40° to the horizontal shaft.

On the upper disc end is formed a shoot leading to a trap-door opening to the inside of the chamber. On the lower disc end is a trap-door opening outwards, and placed so as to close against the lower edge of the cylinder ; the first is the supply or feeding door, the latter the discharging door.

The mixer is generally fixed in the lower floor of a two-story shed ; in the upper floor is a rectangular wrought-iron hopper sufficiently large to contain the ingredients for a charge for the mixer. The bottom of the hopper can be closed by a sliding door, and thence a short shoot guides the materials to the mixer. When the charge has been dropped into the mixer, water is added in due proportion, the upper trap-door closed, and the cylinder rotated for several minutes. Its contents are then discharged into a wagon, or into barrows, for transport.

Messent's Mixer.

153. This is a machine designed and used by Mr. Messent, the engineer to the Tyne Harbour and Breakwater piers, and adopted on many concrete works.

The mixing chamber, of peculiar shape, is of cast iron, mounted on horizontal trunnions supported by a framework, which is sometimes placed on wheels so that the machine can be moved on a tramway. The sides of the chamber are so arranged that the materials, which about half fill the chamber, at every quarter turn are shaken from one side to the other, and are also thrown down from the top to the bottom of the chamber. The materials are thus so thoroughly mingled together that about six to twelve rotations of the chamber, according to the quality of the ingredients, suffice for the thorough mixing of the concrete

Four men can rotate a small machine, larger ones are driven by a steam engine. There is but one door giving access to the interior, it has to serve both as a feeding and discharging door. At Aberdeen South Breakwater, four of Messent's mixers required an 18 horses-power engine, and each mixer turned out about 12 cubic yards of concrete per hour.

OTHER CONCRETE MIXERS.

154. A simple and easily made mixer consists of a timber framework, rectangular in plan, and sheathed inside with planking. The framework may be about 39 inches by 32 inches in plan, and about 8 feet in height; an opening on one side at the foot gives egress to the mixed concrete. On two opposite sides of the interior are fixed inclined planes, which overlap at the middle of the frame and are about 6 to 9 inches apart vertically. There may be as many as eight of these pairs of inclined planes equally spaced.

One of each pair is a fixed plane, the other is capable of being lowered against the vertical side of the framework, and the movable plane is alternately on one side and on the other; the movable planes on each side can be moved simultaneously by means of a lever rod and bell cranks. The planes are all sheathed with sheet-iron. The materials are placed in the framework, the topmost closed planes forming a receptacle; then, by slightly lowering the movable planes, the materials in falling are delivered from one side of the chamber to the other, and are well mixed.

COIGNET'S MIXER FOR "BÉTON COIGNET."

155. This mixer is somewhat similar to the ordinary pug-mill used in tempering clay. It consists of a vertical hollow cylinder of boiler plate resting on a circular base-plate of cast iron. The cylinder is suspended at about 3 feet, or any convenient height above ground, by a strong timber framing housed in a cast-iron jacket bolted on to the cylinder at about 1 foot above the base-plate, which is also held up in place by bolts to the casting. There is thus an unencumbered space round and beneath the cylinder, affording free access for removal of the mixed béton.

In the axis of the cylinder is a vertical wrought-iron shaft, on which are mounted blades or arms, most of them being curved arms set in a spiral round the shaft. The arms of the uppermost set, 3 or 4 in one plane, are broad in the blade, and serve to distribute the materials delivered into the mixer from the hopper. As the shaft rotates the curved arms cut up and intermix the materials. Succeeding the ten curved arms is a set of 3 or 4 helicoidal arms in one plane, by which the materials are stirred and pressed downwards at the same time. The arms of the lowest set are cycloidal, they force the mixed béton out of the cylinder through side apertures.

The béton on passing out of the cylinder is received first on the projecting rim of the base-plate, and thence is forced outwards to fall on to a flat side of the rim of a cast-iron wheel keyed on the vertical shaft. The shaft passes through the base-plate and projects about a foot below, the horizontal wheel rotates with it. From the wheel the béton is swept off at any convenient point by a curved iron scraper fixed to the base-plate of the cylinder, and it may drop on to a trolly or into a wheelbarrow.

The shaft is driven by bevel wheel gearing, mounted at its upper end.

The production of a mixer of ordinary size, about 56 inches high by 22 diameter, is 100 cubic yards daily, an 8 horses-power engine being required to drive it.

Béton Coignet.

156. This is an artificial calcareous sandstone, formed by mixing lime or cement with coarse sharp sand in definite proportions.

The object is to use a large quantity of sand, a small quantity of lime, and water enough for quick assimilation by the lime to prepare it for induration. The lime should be just sufficient, when uniformly diffused, to give a thin coating or envelope to the grains of sand. The ingredients when properly mixed should be barely moist.

Ordinary lime for béton Coignet should be slaked but two or three hours before it is mixed, using as little water as possible; in mixing the béton the water added should be sufficient to render the lime damp and adhesive.

The sand should be coarse grained and quite clean. River sand, from $\frac{1}{20}$ to $\frac{3}{20}$ of an inch in diameter, is best.

Preferably béton Coignet is composed of 16 to 20 volumes of sand, 4 volumes of lime, and 1 to 3

M

volumes of Portland cement; these materials are thoroughly mixed by hand labour on a plank platform, or for large quantities in the form of pug-mill described in Art. 155.

The details of mixing are, alternate layers of sand and lime are thrown into a heap, about a cubic yard in quantity, and are roughly mixed with shovels on the ground (on a platform), the mixture is then passed into the pug-mill. Water is added to the materials when in the pug-mill, a pipe bent to a circle and pierced with small holes is mounted on the upper edge of the hollow cylinder, the supply of water is thus regulated and uniformly distributed. The mixed béton issuing from the base of the cylinder should be slightly moist, and when made into a ball and lightly compressed should retain its form, and harden rapidly.

The béton is to be taken from the mixer and used at once; it should be compressed in the moulds or *in sitû*, by hand ramming, the rammers weighing from 15 to 30 lbs.

Generally $1\frac{2}{3}$ cubic yards of loose béton will compress into a cubic yard. In winter time the ingredients must be heated during mixture, heated air, or steam being introduced in spiral pipes lining the hollow cylinder.

The béton has been used for buildings of all kinds, for stations, houses, a church, for arches, vaults, and retaining walls, for flooring and flagging, foundations for machinery, sewers, aqueducts, water pipes, cisterns, and all kinds of tanks, &c.; for the latter, 5 sand, 1 lime and $\frac{1}{2}$ cement.

BLUE LIAS LIME CONCRETE.

157. Concrete is made of blue lias lime for ordinary foundation work in comparatively dry earth; in damp and in wet earth, and in still fresh water, when the superincumbent load is moderate. For structures in the sea, and in running water, it is safer to use Portland cement, unless a substance containing an excess of soluble silica, such as pozzolana, or similar volcanic ash, be well mixed with the lime.

It is found that blue lias lime concrete sometimes perishes in sea water, it softens to a putty-like consistency and also expands. It is stated by Mr. C. Harrison that at the West Hartlepool Docks and at Sunderland South Docks, blue lias lime was made from Aberthaw pebbles brought to the place of use and there calcined; the mortar was made in the proportions of $1\frac{1}{2}$ lime to $1\frac{1}{2}$ sand and $\frac{1}{2}$ forge cinders, and after 7 years' exposure to sea-water in a dock, became decomposed.

At the Tyne Docks, similar blue lias lime was used, but with the substitution of pozzolana for the forge cinders; the mortar became hard and remained in good condition though exposed to sea-water.

Mr. J. Wolfe Barry insists that blue lias lime must always be thoroughly slaked before use. He states that at the Barry Docks, the Aberthaw lime from the kilns was completely wetted on the floor of a shed, and was then covered with sand, and left for not less than seven or more than fourteen days before being used as a mortar, or for concrete. Some masonry was built with lias lime mortar, mixed as dry as possible, the lime being previously kept in dry covered sheds. The mortar of the blocks of masonry set perfectly hard, but in a few weeks the blocks burst to pieces. In another instance of the use of an imperfectly slaked lias lime, large blocks of masonry were lifted completely off their beds by the expansion of the mortar in the joints. At the Barry Docks, the mortar used was made of 1 lime to $1\frac{1}{2}$ sand and $\frac{1}{4}$ of hard burnt clay, ground in a mortar mill with plenty of water. It is not advisable to add clay to lime in making mortar, unless the clay be rich in gelatinous silica, and be thoroughly burnt till the silica is in a condition to combine readily with the caustic lime.

In preparing lias lime concrete, after the thorough slaking of the lime, the sand, lime, and water should be mixed to make a creamy paste, the proportions being 1 lime to 2 of sand. Then the clean gravel pebbles, or broken stone, say 4 to 6 parts, to be added, and to be very thoroughly mixed by turning over the whole mass with shovels six to seven times. The mixed concrete to be deposited in 9-inch layers, each layer to be well rammed, till the mortar shows flushing up to the surface, and if there appears to be any deficiency of mortar, if all the spaces between the stones are not filled, each layer is to be grouted with additional mortar.

At Naples the proportions used for a lime concrete are stated to be—1 white lime (probably a rich lime), 2 of pozzolana, and 3 of volcanic scoriæ; the whole thoroughly mixed, and left for some days (not exceeding fourteen) for the lime to slake completely and to act on the soluble silica, before being used.

At the Avonmouth Docks, a lias lime concrete was used of the composition 1 lime, 2 sand, 2 furnace ashes, 2 of broken stone, or 1 to 6 concrete; this was used beneath a superstructure of cement concrete.

Burnell gives proportions in bulk recommended by General Treussart: 20 per cent. hydraulic lime to 20 of trass, 20 of sand, 13 gravel pebbles, and 27 of broken stones; or 20 per cent. of hydraulic lime, 28 of pozzolana, 14 of sand, and 38 of broken stone; both these are for concrete in water. First a mortar is made, then pebbles and stones are thoroughly incorporated. It is stated that the second concrete should not be used for about 2 hours after mixing.

A proportion used by Vicat in a bridge is given as 20 per cent. hydraulic lime in paste, 30 of granitic sand, and 50 of gravel pebbles. The concrete was about one-third less in volume than the sum of the ingredients.

CHAPTER VIII.

CEMENT CONCRETE IN SEA-WATER; DEPOSITION OF CONCRETE; WEIGHT OF CONCRETE; MEMORANDA.

Cement Concrete in Sea-Water.

158. SOME doubt has lately been thrown on the permanence of cement concrete exposed to the action of sea-water, especially when the concrete is alternately submerged and exposed to drying in air, and also subjected to great differences of hydrostatic pressure, due to a considerable depth of water being at times retained on one side only of the concrete wall, and the mass being sufficiently permeable to permit percolation.

159. In a case of this kind (described in Proc. Inst. C.E., vol. cvii.), some concrete was found to be disintegrated and readily washed away, and a comparatively large amount of hydrate of magnesia had been deposited among the softened and decomposed concrete. It is stated that little was left but sand and gravel, and a creamy substance consisting of 80 per cent. of hydrate of magnesia.

The only analysis of the cement used for this concrete was obtained from a test briquette which had been preserved; none of the cement powder was procurable. The ingredients were stated to be :—

Caustic lime	45·39
Carbonate of lime	8·18
Hydrate of lime	11·26
Silica	20·92
Alumina and iron oxide	13·10
Sulphuric acid	0·82
Magnesia	0·33

The cement was supplied subject to a specification of tests :—of sieving, to pass through a gauze sieve of 1000 holes to the square inch, leaving not more than 5 per cent. residue ; of weight, to be not less than 115 lbs. and not more than 124 lbs. to the imperial striked bushel ; and neat cement briquettes, after being in air twelve hours immediately after moulding, and then immersed in water for seven clear days, to bear an average tensile dead weight of 1000 lbs. avoirdupois on a section of $1\frac{1}{2}$ by $1\frac{1}{2}$ inch, the minimum dead weight being 750 lbs.

The concrete was made in different proportions ; about one-half was made of 1 cement to 3 sand, 3 of ballast, and 6 of large stones of over 40 lbs. weight, the other half of 1 cement to 4 of sand, 4 of ballast, and 8 of large stones. The large stones were not mixed with the other ingredients, but were placed by hand on the deposited concrete to be grouted —that is, the sand and ballast concrete was filled in round them.

160. The analysis of the cement briquette leads to the opinion that the cement clinker was underlimed, or

underburned, or both, or had deteriorated from exposure; and the tests imposed were almost valueless, the sieve test being too coarse, the weight test of little use, and the neat cement briquette test liable to be misleading.

The composition of the concrete is defective in the excess of sand used in both kinds, the result being to produce a weak and permeable concrete. Mr. Messent states, as the result of his extensive experience in pier, harbour, and dock work, that a good proportion for a non-permeable concrete in sea-water is 1 cement to 2 sand, and 5 of stones; or, 1 cement to $1\frac{3}{4}$ sand, and $5\frac{1}{2}$ stones; the mixed sand and cement to form about one-third of the whole mass, and about half of the bulk of the stones.

161. Sea-water generally contains about $3 \cdot 2$ parts of chloride of magnesium, $2 \cdot 1$ parts of sulphate of magnesium, sulphate of calcium $1 \cdot 4$, chloride of sodium $28 \cdot 0$, and small quantities of bicarbonate of calcium and bromide of magnesium, and about 964 parts of water out of 1000 parts; the composition varies in different seas.

The compounds formed by the reactions of the oxides of aluminium and iron with lime are easily softened by pure water, while the silicate of lime remains hard. The magnesium sulphate and chloride in sea-water produce corresponding compounds of lime in permeable concrete, the calcium sulphate attacks the actively cementitious tricalcium silicate and aluminate of the cement, and forms crystalline compounds.

The sulphates are the most dangerous ingredients

in sea-water. They act upon aluminium and iron compounds of lime, which are rapidly softened and decomposed. Sulphates acting on hardened cement, form sulphate of lime, mixed with aluminium and iron compounds, which crystallise and expand, thus destroying the cohesion of the mass, and causing cracks, fissures, and bulges in the concrete. Hydrated lime (generally to be found in set concrete) displaces the magnesium salts in sea-water from a state of solution, the lime is converted into soluble calcium salts, and the magnesia is precipitated as a soft bulky hydrate, which is devoid of cementitious properties, and is inert; it may clog and choke up the pores of the concrete, but does not cause any rupture by expansion; in fact, the separation of the hydrate of magnesia is the visible, but innocuous, sign of the before-mentioned action of the sulphates.

Mr. Draper states that the space occupied by one molecule of calcium hydrate, as compared with that of one molecule of magnesium hydrate, is as $34 \cdot 8$ to $35 \cdot 14$—a difference so slight that the disintegration of concrete cannot be due to the change; also, that the calcium sulphate deposited in concrete from mere exposure to sea-water, or from immersion without percolation, cannot be in sufficient quantity to act injuriously, although the action ascribed to the sulphates is the disruptive force of the process of crystallisation. In some cases there may be as much as 4 per cent. of calcium sulphate in the cement as produced from the kiln.

162. It is found that a solution of gypsum destroys Portland cement, probably by the formation of sul-

phate of lime, which crystallises with increase of bulk ; the iron salts also form a double compound, a white voluminous salt ; similarly, alumina with sulphate of lime forms a double salt ; the aluminium compound crystallises when in contact with a large quantity of water, and increases in volume.

The salts in sea-water are found to attack most readily cement concrete which is permeable or fissured, and through which the sea-water percolates freely ; the hydrated silicate and aluminate of lime are attacked, and the injury is proportional to the poorness of the concrete in silica, and its richness in lime, alumina, and iron oxide.

163. Chemists consider that Portland cement, when indurated, becomes a tricalcic silicate and a tricalcic aluminate, having a composition of CaO 63·08, SiO_2 16·89, Al_2O_3 9·57. The average of three analyses of good cement gave CaO 61·02, SiO_2 20·29, Al_2O_3 8·25, showing a close approximation to the theoretical composition, excepting as regards the silica.

PRECAUTIONS TO BE OBSERVED IN THE USE OF
PORTLAND CEMENT CONCRETE IN SEA-WATER.

164. Portland cement concrete for use in sea-water must be mixed in suitable proportions, the cement to be as finely ground as practicable, to be tested in the form of mortar briquettes, which are to be subjected to immersion in boiling water, to show a gradually increasing resistance to tensile stress at 7, 14, and

28 days after moulding, and within one month after grinding to be of a specific gravity ranging from 3·05 to 3·1, every possible precaution being taken to prevent the supply of overlimed and underburned cement. The materials for mixing to be washed clean, free from all dust, dirt, and clayey matter; the pieces of gravel, shingle, broken stone, &c., to be of all sizes, from about half an inch diameter, to about 2½ inches. If the concrete is not required to be impervious (to the extent of preventing percolation when a wall sustains on one side only a considerable height of water), sand need not be used to fill up the interstices between the particles, except for the outer 6, 9, or 12 inches of the faces of large blocks and of thick walls; this facing it is well to make as dense as possible in all cases of concrete in sea-water. A dense concrete wall may be made of 1 cement, 2 of sand, and 4 of stones, with larger blocks of stone embedded in the hearting, the outer facework being 1 of cement to 2 of sand, for 9 or 12 inches thickness and upwards.

For this facing, and for dense concrete to resist percolation, sand should be used to fill up the spaces between the stones, and the sand should vary from fine to coarse, or from about one twenty-fourth of an inch diameter up to one-eighth of an inch, the proportion of the finer particles being about one-third. Cement must be used in proportion to the total area of surfaces to be coated—that is, in larger proportion when sand is used, and the finer the sand the more cement. Coarse sands are liable to make permeable mortars; fine sands make porous mortars, but less permeable than coarse. A proportion of 1 cement to 2 of fine sand will probably be impermeable.

165. The concrete must be thoroughly mixed, and must be deposited in place without any serious disturbance of the mixing or loss of cement. When a layer of about 12 inches thickness has been deposited, if not submerged it must be compressed by ramming till the mass becomes pulpy. This compression of the concrete is of considerable value, and the drawbacks to the deposition of newly-mixed concrete in deep water to form a monolithic mass, consist in the difficulty, and in many cases the impracticability, of applying this compression, also of preventing a serious loss of cement. To obviate this, concrete is sometimes used in large blocks, which are moulded in a moulding yard with the aid of the most complete apparatus for mixing and moulding. For impermeable walls of great thickness, the central portion may not require to be made of the densest concrete, but may be more cheaply made of large clean stones forced down into a freshly-deposited layer of sandless concrete. The large stones should be placed not touching one another by about 3 inches, but with ends overlapping. The outer faces of such a wall should be made of dense impervious concrete to a depth of 6 inches, 1, 2, or 3 feet, or more, according to circumstances.

166. There is no adequate cause for doubt as to the durability of concrete made of finely-ground cement of good quality, well mixed in suitable proportions, with sufficient water for complete hydration, and either properly moulded in blocks or deposited under

all requisite precautions to form a monolithic mass. There are numerous instances of durability for periods up to 40 years' use. Mr. H. E. Jones, of the Commercial Gas Company, cites an instance of the successful use of cement concrete tanks for ammoniacal solutions stronger than sea water; the inner surfaces being coated with cement plaster only, which was hand-floated—that is, worked with a trowel to a perfectly smooth hard watertight skin, like an enamelled surface. (See Proc. Inst. C.E., vol. cvii. p. 125.)*

167. In certain works what is termed "plastic" concrete has been used, that is, the concrete has been mixed and allowed to begin to set, then it is again broken up, and deposited in place, being well compressed by ramming. This "plastic" method is considered by some engineers to be a mistake, but the system is advocated by Mr. Kinipple in a paper on "Concrete Work under Water," in Proc. Inst. C.E., vol. lxxxvii., where examples of its successful use are mentioned. It appears to be a method which may be successful under very careful supervision, but is not one which can be generally adopted, as the time during which the setting concrete may be broken up is limited, and must not be overpassed.

168. The use of clean sea-sand and clean sea-water for mixing is probably advantageous for concrete works exposed to a hot dry climate; the concrete

---

* Mr. Jones, engineer and manager to the gas company, has kindly furnished the proportions of the plaster; half an inch thickness of 1 cement to 1 sand, covered with one-quarter of an inch of neat cement.

will retain moisture in its pores, and not be so liable to lose its water of hydration. Concrete exposed to severe drying should be occasionally flooded with water.

169. Dr. Michaelis, in a paper published in Proc. Inst. C.E., vol. cvii. p. 370, suggests the following remedial measures in the case of concrete that is beginning to show signs of disintegration:—Wash the outside of the concrete with bicarbonate of ammonia and allow to dry; this converts the lime of the surface into carbonate, which is not liable to be attacked by sulphates. Another remedy is to coat with magnesium fluosilicate, which closes the pores of the concrete, and converts the free lime into fluoride and silicate of calcium. Another suggestion is to dissolve about 2 to 3 per cent. (of the weight of the cement) of barium chloride in the water used for mixing the concrete; the result is the formation of insoluble barium sulphate with the sulphates of the sea-water, while the magnesia remains in solution as magnesium chloride. A coating of soluble silicate of soda syrup is one of the best preservatives, by reacting on the lime and forming a silicate.

170. It is interesting to note that in Burnell's excellent rudimentary treatise on limes and cements, published nearly thirty years ago, the cause of failure of limes and cements in sea-water is stated as investigated by Vicat, with identically the same results as were found recently—viz. that the hydrates of lime were converted into soluble chlorides, and that magnesia was deposited in the vacancies; but it was not then recognised that the deposited magnesia is inert.

Full prominence is, however, given to the formation of calcium sulphate from the action of calcium hydrate in the mortar or concrete on the magnesium sulphate in sea-water; and the expansion, due to crystallisation, of the calcium sulphate is recognised as a means of destroying cohesion, and thereby leading to disintegration.

The Deposition of Concrete.

171. Concrete is in general use as a first layer, to cover a foundation bed consisting of more or less compressible soil, or to make good to a level surface the irregularities of a hard bed. It is generally cheaper than brickwork, and an intelligent labourer can be quickly instructed in the best method of forming a firm solid platform of concrete to carry the masonry superstructure.

172. As concrete has no great amount of resistance to transverse bending, Sir B. Baker states that for a concrete foundation bed resting on plastic clay, the thickness of the concrete should not be less than about $1\frac{3}{4}$ times its projection beyond the face of a pier or wall, or beyond the edge of masonry footing courses; or its projection to be four-sevenths of its thickness. (Proc. Inst. C.E., vol. lxv.)

173. Mixed concrete should not be tipped into a foundation trench from a greater height than about 5 feet; if possible it should be wheeled in, or be lowered in barrows or boxes by a crane or hoist. If it be necessary to shoot the concrete down an inclined plane into the trench, or if tipping from a height be adopted, the

proper mixture of the ingredients may be disturbed, the larger stones separate from the smaller particles, and each tipped mass assumes a conical shape, comprising a solid dense hearting, mainly of the finer particles, and comparatively loose slopes of large-sized materials. To obviate this detrimental result, the "tipped" or "shot" concrete should be remixed at once by spadework in the trench; but owing to difficulties of supervision of this essential precaution, it is better not to allow the practice of tipping, &c.

174. The concrete should be deposited in layers not exceeding 18 inches in thickness for ordinary foundation work, and in 12-inch layers for important work, and each layer is to be carried uniformly over the whole area of the foundation, or of the structure, so that no part is ever more than one layer higher or lower than the rest. Each layer should be well packed into the foundation trench, or into the mould-frame, first by "packing," or cutting and cross-cutting into the layer with spadework, and then ramming with wooden or metal rammers weighing at least 15 lbs. per man. The surface of the concrete, when sufficiently rammed, appears to be moist with the cementing material. The ramming and packing must be done as soon as the concrete is deposited in place, and be completed before the cement has begun to set. Rammed concrete exhibits a gain of about 23 per cent. in strength over not-rammed, and there is an important gain in density and imperviousness. For walls of concrete built in frames, spade-packing and light wooden rammers are to be used.

175. It is a bad practice to allow concrete to be

carried up to a considerable height in short lengths, as the settlement of contiguous lengths will not be uniform, and divisions and cracks in the concrete will be developed, unless special precautions are taken.

176. The separate layers, in concrete intended to be water-tight, must be thoroughly incorporated together; if the under layer has not yet set, the intermixing may be effected by spadework, cutting through the top into the underneath layer, and working the spade backwards and forwards.

177. Cases occur where the concrete must of necessity be carried up in short lengths, or where it has to be for a time omitted from some part of the work. With some cements there appears on the surface of the setting concrete a scum (*laitance*), which generally consists of alumina, silica, and some lime. This scum must be removed if the concrete is to be sound and impervious, with the successive layers perfectly cemented together. When the new concrete is to be deposited, the surface of the old work must be well watered, swept clean, and, if necessary, scrubbed with wire brushes; in some cases the old surface is partially, or wholly, removed, and a roughened surface made by pecking with a sharp-pointed pickaxe. The surface thus cleansed and prepared is then lightly dusted over with neat cement just before the new concrete is deposited.

For a temporary gap in a length of concreting, the ends of the layers on each side of the gap should be left stepped back about 12 inches each layer.

A case of failure in a concrete reservoir wall was

traced to the neglect of removal of scum from the surface of setting concrete. The scum was attributed to the use of dirty water, but some of it probably came from the cement.

178. All concrete expands and contracts more or less with alternations of heat and cold, and there is also a contraction with age. In an experiment on a prism of neat cement 7 inches long, the contraction in a year was found to amount to 0·162 to 0·113 per cent., according to the dryness of the situation. Another experiment gave a contraction of neat cement of 0·04 per cent. in one month, continuing thereafter at a reduced rate.

179. Concrete paving in the open air should be put down in cold weather, short of actual freezing during deposition; it may be covered with straw or other material to guard against occasional frost until thoroughly hard. As the ramming of a thin horizontal layer of cement or concrete may be impracticable, a good method is to fix firmly at the sides of the layer (and enclosing a convenient area) wooden straight-edges, with the upper edge level with the surface of the intended layer. Fill in the cement concrete, or rendering, and then gently ram the surface down with the straight edge of a vertical board (say $1\frac{1}{2}$ inch thick) resting at one end on the wooden side of the enclosure ; after chopping down, the surface may be strickled, made plane and true, by dragging the board, resting on opposite wooden sides, from one end to the other of the temporary enclosure.

180. For vertical surfaces, such as the face of a concrete wall rendered in finer stuff, the best method is

to deposit the finer stuff against the mould framing, inside a thin temporary partition of sheet iron or of thin hard wood, then the coarser concrete; and after removing the partition, chop the coarser stuff with a spade till it is incorporated with the fine stuff densely packed close up to the frame. The thickness of the coating, or rendering, for house walls, &c., may be 1 to 1½ inch.

181. A layer of fine stuff, forming a smooth surface to coarser concrete, is sometimes found to blow up and shell off. The cause of this is partly the swelling of too new, or of inferior, cement; partly the dryness of the under stuff when the fine coating is put on; and sometimes the workman uses too little cement in the coating, and he works up the major part to the surface, leaving a deficiency underneath. The coarse stuff in such case may be 1 to 6 concrete, and the finishing coat ½ inch to 1 inch thick of 1 cement to 2 of good sand of mixed sizes of grains.

182. In concrete retaining walls a vertical joint of ½ inch width, and penetrating 6 to 12 inches into the facework (generally the whole depth of the dense face concrete) is formed at every 20 or 30 feet length of wall, in order to counteract the tendency concrete often shows to crack during the first six or twelve months. When the concrete has ceased to contract to any material extent, this joint can be filled with cement mortar or with neat cement.

In paving slabs an open joint about ⅜ inch wide is often provided for every 100 feet superficial; and sometimes an open joint is provided for an arch ring by temporarily inserting a wooden plate at each springing, and then the whole of the upper surface of the arch and its supports is to be covered with an impervious layer of asphalte.

183. The surface of concrete flooring must be protected by a covering of planking, or of sand, or of sawdust, till it is hard enough to bear traffic. For ordinary footways a layer of broken stone or broken brick, 4 to 6 inches thick (or more if the foundation bed be soft), must be laid and well rammed till it is quite compacted (the foundation bed must be well drained), and then the finer concrete, laid down 2 to 3 inches thick, and 3 to 4 inches for a cart traffic, is thoroughly consolidated; it must not be worked to a very smooth surface, and for horse traffic is slightly ridged or indented to give better foothold. The broken stone or brick layer must be well watered before the concrete is put down. The largest sized stone in such concrete generally measures not more than half an inch in any direction, and broken granite or any hard igneous, or hard crystalline, rock answers well. One cement to 1 sand and 3 of stones is a useful proportion for slabs; for blocks, such as steps, 1 cement to 7 stones may be used for the interior of the block. Crushed fire-brick, blue Staffordshire brick, or brindled bricks, or hard-burned semi-vitrified common bricks may be used.

For a floor for ordinary purposes, a concrete of 1 cement to 2 of sand (coarse to fine), and 3 of broken stone, or small pebbles not larger than an inch across, will give good results. For a foot pavement to resist abrasion, the best results are from an equal admixture of quartz or granitic sand and cement; nearly as good results are, however, obtained from 1 cement to 2 sand.

184. In concrete walling for houses the usual thickness of wall for a two-story house (24 feet high) is 9 inches, and for every additional 6 feet to the height of the wall add 3 inches to the thickness of the lower 24 feet of walling. (This subject has been mentioned in Art. 144.) Concrete walls, when well made, may be thinner than corresponding brick walls by onefifth. Arches in concrete should not be built in ring layers but in continuous voussoir sections with temporary radial joints, using wooden boards or metal sheets as divisions; these voussoir sections to be from 12 to 18 inches in width, and their depth the full thickness of the arch ring. Suspended beams, floors, and flat or arched roofs, and all similar constructions, must be rigidly supported or housed at the ends of the beam, or at the sides of the span, so that there shall be no yielding or spreading apart of the supports.

185. It is difficult to estimate the strength of concrete used in beams and arches, as the quality of the cement, the qualities of the other ingredients of the concrete, the mixing and deposition, all have material influence on the strength of the concrete, and therefore no trustworthy data are yet available for the calculation of the strength of concrete under so many varying conditions.

As an instance of concrete arching for a span of 11 feet, rising $1\frac{1}{4}$ inch per foot of half-span, 5 inches thick at the crown, and 9 inches at the haunch; 1 cement to 1 sand to 5 of broken hard brick. At

28 days this arch carried safely a distributed load of 3 cwt. per square foot.

An article on the construction of concrete quarters for native railway officials in India is given in Proc. Inst. C.E., vol. ci. p. 186 ; and one on concrete building at Simla, India, in vol. lxxxiii. p. 390.

186. Concrete is sometimes deposited in still water through a wooden, iron, or canvas tube, lowered till it almost rests on the surface of the submerged ground. But there is always more or less loss of cement in dropping mixed concrete through water, and a better plan is to use a skip or discharging-box, which can be filled with mixed concrete, closed, lowered through the water to the bottom, and then opened to discharge its contents. These skips can be made of wood, or of wrought iron and steel ; the latter are the better. The metal skip-box is made in two flaps, forming when closed a semicircular sectioned box with the diameter (or flat side) uppermost ; each flap is therefore a quadrant of a circle. The external hinge bars, forming part of the framing of the box, cross at the hinge-joint and are prolonged past the joint upwards, and the suspension chains are attached to these two projecting horns on each side of the box. The weight of the material in the box brings additional stress on the chains or links attached to the horns to force them together, and thus keeps the box tightly closed when suspended. As soon as the box is lowered to the bottom, and ceases to hang from the suspension chains, it is opened by pulling opening chains, one attached to the outside of each half of the box.

187. In engineering work concrete is deposited gradually to form practically monolithic masses of any size ; or it is placed in wooden or iron moulds or frames, to form blocks, which are allowed to harden, and then are transported to the work, and slung and lowered into place from powerful cranes. The limit to the size of the moulded block is the power available for handling and depositing, and 30- and 50-ton block cranes are in common use.

188. Sometimes larger masses are required for the construction of piers and breakwaters in the sea, and they are generally moulded in the wells of large hopper-barges. A bag is made of two or three thicknesses of jute sacking, the well of a hopper barge being lined with the sacking to receive the mixed concrete ; or the block is moulded, and then is supported between pontoon barges, and thus can be carried to the place of deposition and there lowered or dropped into its place. Concrete in bags has been made in blocks of 100 tons and upwards, but the objection sometimes raised against the use of bags is that the sacking effectually prevents any adhesion between the blocks. Still, for the construction of a broad platform of concrete covering a soft sea bed liable to be scoured away, the concrete in bags is very serviceable, and concrete in bags can be used to bring to an approximate level surface any rugged foundation bed, or to form a protecting apron round the base of a structure exposed to the wash of tides and sea currents.

Bags are sometimes used to protect small masses of concrete from running water. Ordinary sacks are filled with mixed concrete, and then are piled up to form a protecting face wall where a river bank is being scoured away; the bags prevent the washing away of the cement before the concrete sets.

Where strong springs are met in foundation trenches, or where strong currents of water have to be dealt with, concrete is sometimes deposited inside tarpaulin, or waterproofed sacking.

189. In the so-called monolithic work, the form to be given to the mass is outlined by timber framing lined with planking, which is coated over with soap solution or with a wash of clay mud, to prevent the concrete sticking. An advantage of this method of construction is that only simple appliances are required, the framework is re-used, and a small crane only is required to lift and lower the mixed concrete in discharging boxes. This compares very favourably with the expensive gantries, traveller cranes, staging, &c., required for making, storing, and handling large blocks, transporting, and setting them in place.

No framing, however, could long withstand the shocks of heavy seas; while block-setting, from powerful overhanging cranes, can be carried out in moderately rough seas, and in a greater depth of water than could be reached by framework; and, in case of sudden storms, the cranes, &c., can be run back along the tramway to a place of safety.

190. Another important use of concrete is in the construction of hollow cylinders, or of tubes of various sections to be used in obtaining solid foundations for piers and abutments of bridges, quay and dock walls, and engineering structures generally; also for machinery shops, warehouses, &c.

The large tubes are moulded in lengths of about 30 inches each, and are built-up vertically, cemented together *in sitû*, so as to become practically monolithic. They are easily sunk to a great depth in soft soils, by excavating the soil in the interior (or shaft of the tube) and undermining its edge, which is generally tapered from the inside to the outside of the shell of the tube, and is also protected with, or formed altogether in, metal (cast or wrought iron). At the same time as the undermining (misering), heavy loads are placed on the top of the tube to force it down steadily into the ground. The excavation in the interior is carried on by manual labour, or by specially designed excavating machines. The tubes are generally termed "wells" when used in foundation work; but the same method of construction and of sinking is adopted with concrete chambers, sometimes open-topped, sometimes closed over to be air-tight, called caissons, used in laying bare for inspection and construction, foundation beds at sites covered with a considerable depth of water.

A valuable series of short papers on concrete works is to be found in Proc. Inst. C.E., vol. lxxxvii., among others; also in the engineering periodicals.

191. The weight of Portland cement concrete depends on the nature of the ingredients, the quantity of sand and cement used, and the extent of compression applied to the deposited concrete.

All kinds of natural stones vary widely in their density; the cement and siliceous sand may be considered as fairly constant; the following weights must therefore be considered as approximate only, per cubic foot :—

Portland limestone concrete, about 125 to 130 lbs.
Granite concrete .. .. .. „ 135 to 140 „
Slag „ .. .. .. „ 110 to 118 „
Ballast „ .. .. .. „ 130 to 140 „

Ballast concrete will vary widely in weight, according to the quantity of sand used. Ordinary gravel weighs about 2900 lbs. per cubic yard of 21 bushels.

If the concrete be compressed by ramming *in sitû*, these weights may be increased by about 4 to 15 per cent.

The lightest concrete is made of crushed coke, and it may weigh about 70 to 80 lbs. per cubic foot. Semi-vitrified burnt clay concrete will weigh about 110 to 115 lbs., broken brick concrete from 110 to 120 lbs. per cubic foot.

The effective weight of concrete deposited in water is its weight in air diminished by the weight of the volume of water displaced; and it is necessary to use blocks of large size and weight in work exposed to strong currents and to the shocks of heavy waves, a dovetail or interlocking pattern of jointing may also be used. In such work cavities and open joints should be avoided, as the air in the cavities may be strongly compressed by the impact of waves, and its subsequent sudden expansion has been known to dislodge blocks and cause extensive injury to the structure.

USEFUL MEMORANDA.

1 lin. inch	= 25·3998 mm.
39·3704 lin. in.	= 1 metre = 3·2809 ft.
1 sq. in.	= 6·4516 sq. cm.
10·7643 sq. ft.	= 1 sq. metre = 1·196033 sq. yd.
1 sq. foot	= 0·09289 sq. metre.
1 bushel	= 1·283 cu. ft. nearly = 36·328 litres nearly.
1 cu. yd.	= 21·046 imp. striked bushels = 0·7645 cu. metre.
1 cu. metre	= 35·3157 cu. ft. = 1·3079 cu. yd.
1 kilog.	= 2·2046 lbs. (avoir.).
1 lb. avoir.	= 16 oz. = 7000 gr. (avoir. or troy).
1 cwt.	= 50·802 kilog.
1 litre	= $\frac{1}{1000}$ cu. metre = 0·220215 gallon.
1 litre	= 61·027 cu. in. = 0·027527 bushel.
1 hectolitre	= 2·7527 bushels, or 22·02 gallons.
100 lbs. per bushel	= 1248 grm. per litre.
1 kilog. per sq. cm.	= 14·223 lbs. per sq. in.
1 lb. per sq. ft.	= 4·8826 kilog. per sq. metre.
1 lb. per cu. ft.	= 16·019 kilog. per cu. metre.
1 kilog. per cu. metre	= 0·062425 lb. per cu. foot.
900 per sq. cm.	= 5806 per sq. inch.
5000 ,, ,,	= 32,257 ,, ,,

APPENDIX.

MANUFACTURE OF KANKAR LIME.

By J. BENTON, F.C.H.,

Executive Engineer, Bari Doab Canal, India.

THE lime chiefly used in the plains of the North of India for engineering purposes, is made by calcining kankar with charcoal. Kankar is a calcareous concretionary deposit formed in the alluvial soil of the plains of India, from carbonate of lime held in solution. It is composed chiefly of carbonate of lime, silicate of alumina, and silica in grains. The kankar best suited for lime-making contains a high proportion of carbonate of lime and a small proportion of silica in grains; it has a dark brown or dark blue colour on the surface of a fresh fracture, is even and close in texture, and should be rather free from glistening particles. Kankar occurs either in beds of irregular nodules, varying from $\frac{1}{4}$ inch to 4 inches, but usually about $1\frac{1}{2}$ to $2\frac{1}{2}$ inches across, or is found in continuous layers, varying from 6 inches to 18 inches in thickness, and then has to be quarried like stone, and is termed block kankar.

In general, nodular kankar on being dug is sufficiently cleared of earth by beating it with short stout sticks; in certain cases greater purity is attained by washing in clean water. Nodules exceeding 2 inches across should be broken, and block kankar should be broken for lime-making into pieces about 2 inches across.

The kilns in which the kankar is burned are shown in Figs. 1 and 2; they are built of underburnt or of sun-dried bricks, laid in mud mortar, or may be made of mud walling. The outer and inner surfaces of the kiln should be plastered with clay, in order to assist the retention of heat. For the supply of lime to an isolated large work requiring a series of fresh supplies at short intervals of time, a group of triple or quadruple kilns, built to No. 1 pattern, is well suited, enabling the production of lime to be carried on in all stages simultaneously, one kiln being in course of charging, one being fired, another cooling and in course of unloading. Three or four such kilns make a convenient group, and for a large work there may be one or more groups.

If a large quantity of lime is required to be ready for use at a certain time, a number of large kilns, as in Pattern No. 2, will be found suitable; and for a small work of similar character one of these kilns will answer. An economy of fuel attends the use of large kilns.

The fuel in general use is charcoal, made by burning any hard wood in pits dug in the ground, the wood pile being covered over with a layer of earth to prevent the escape of flame. Native contractors are generally able to supply charcoal of good quality; that from soft woods is, weight for weight, inferior in heating power to the hard-wood charcoal. The weight of a good charcoal is about $20\frac{1}{2}$ lbs. per cubic foot, that of soft-wood charcoal is less; a short experience in the inspection of hard-wood charcoal enables an engineer to judge its quality approximately.

Kilns—Air-flues.—To ensure quick and uniform burning of nodular kankar, air-flues must be formed in the floor of the kiln, and be continued upwards through the mixed kankar and charcoal filling the kiln. These flues are indicated by dotted lines in Figs. 1 and 2, and are shown in detail in Fig. 1 B. As a general rule, the floor and vertical flues are planned so that no part of the filling is more than 6 feet distant from a flue in any horizontal section of the kiln. But the planning of the flues may be modified according to the character and size of the pieces of the kankar to be burned; if it be of good size, say 2 inches across, with practically no smaller fragments, as is the case with broken block kankar, vertical flues are not absolutely necessary. The vertical flues are best formed of bricks laid dry and close together, as shown in Fig. 1 B; the products of combustion pass up through the filling of the kiln, adjacent to the four faces of the brick pillar.

Loading or Charging the Kiln.—The floor of the kiln is first covered with a course of *upla patti* (dried cow-dung cakes) laid on edge, a fuel readily obtainable in any Indian village. This floor layer is required to ensure nearly simultaneous firing throughout the area of the base of the filling. On the top of the *upla patti* a layer of charcoal, 1½ inches deep, is uniformly spread. Then ignited pieces of *upla* are inserted in the *upla* floor layer, at intervals of about 2 feet apart; and the charging of the kiln is at once begun. The charge of kankar and charcoal is mixed on a brick floor; a measured quantity of clean kankar, say 100 cubic feet, is spread out in a layer of uniform depth of 6 inches, and charcoal in the proportion of—

(1) 44 cubic feet charcoal per 100 cubic feet kankar in the hot weather (April to the end of June, the kankar being fairly dry);

(2) 50 cubic feet of charcoal per 100 cubic feet of kankar in the cold weather (when the kankar may be damp, or have been recently exposed to rain, or to wetting)—this proportion may hold good from July to the end of March, according to the character of the season;

is evenly spread over the kankar layer.

The kankar and charcoal layers are then filled, mixed together, into baskets, carried to the kiln, and deposited in horizontal layers over the inch and a half layer of charcoal. The fire in the upla and charcoal layer below does not spread very rapidly, and there is no difficulty in completing the charging of the kiln in one operation if the proper supply of cleaned kankar and charcoal be at hand. The kankar is generally stacked in heaps conveniently near the mixing floor, and if the pieces are at all damp, the heaps should be spread out to dry for a day or two before burning. Damp or wet kankar will probably come out of the

MANUFACTURE OF KANKAR LIME. 193

kiln underburnt, if only the specified amount of fuel be used.

Duration of Burning.—Small kilns charged with broken block kankar will burn out in two days, while large ones charged with small earthy nodular kankar will take from three to five days in burning. No covering is placed on the top of the kiln, the charge is heaped up above the top of the wall of the kiln in a hemispherical mound, which sinks as the charcoal is consumed.

Good charcoal leaves very little ash, an important advantage attending its use as a fuel for lime burning; it also burns very evenly, and the out-turn from a well-managed charcoal-fired kiln is uniformly burnt. A well-burnt piece of kankar can be broken somewhat easily by hand.

Grinding or Pounding.—On large works the calcined kankar is reduced to fine powder in disintegrators. These machines give finest reduction and consequently best results. When the disintegrating mills are not available, the calcined stone is ground under edge runners of stone or hard cast iron, running in a circular track; and, failing edge runners, the stone is pounded by hand under hard wood mallets, or iron hammers. The finer the reduction the better is the lime for mortar mixing.

Deterioration of Lime.—All kinds of quick or caustic lime rapidly deteriorate if kept for any considerable length of time, even in the dry season in India, and kankar lime should always be used when newly burnt. Stale lime sets slowly, and never attains great ultimate strength. The limit of keeping

O

after the burning should be about one or two weeks in the wet season, and two or three months in the dry season; sometimes the dry weather term is limited to one month after burning. Kankar lime has been stored most carefully, and covered over with plaster to preserve it uninjured during the rainy season (June 20 to September 30), but damp air has gained access, and has rendered the lime untrustworthy.

Admixture of "Surkhi."—Well-burnt brick, when pounded or ground to powder, is known as "surkhi," and is used in India as a substitute for sand in the preparation of mortar. Sand is but rarely to be procured of suitable quality, and free from clayey matter; it is generally loamy, or is in fine rounded grains having smooth texture of surface. The brick for "surkhi" should be made from an unctuous clay, and be as free as possible from sand (silica in grains), but should contain a large proportion of soluble silica, which will be acted upon by, and will combine with, the caustic lime, forming a durable silicate of lime. Nodular kankar usually carries some clayey matter which has not been completely knocked or washed off, and the calcined nodules have then a certain amount of burnt clayey matter adhering, sometimes sufficient to render unnecessary the addition of "surkhi." As the amount and quality of the adherent clay must be uncertain, it is better for important work to calcine clean nodules or lumps, and add a measured quantity of "surkhi" of approved quality. The amount of burnt clay thus added depends upon the proportion of soluble silica contained in it, and may

MANUFACTURE OF KANKAR LIME.

advisedly be made the subject of experiment in each case of employment. To a rich, or comparatively pure, block kankar lime, "surkhi" may be added in the proportion of $\frac{1}{2}$ to 1 of "surkhi" to 1 of lime, by bulk. If, however, the "surkhi" be from sandy bricks only, not containing a considerable proportion of soluble silica, the addition will probably cause a lessening of the strength of the unmixed lime—in fact, will be an adulteration of the lime with a quantity of nearly inert matter.

The materials for "surkhi" must be carefully selected by the engineer, preferably from the Government stock; and only selected and approved materials allowed to be crushed to powder. The workmen, if left to themselves, will naturally take soft and underburned bricks, &c. When "surkhi" is supplied by contractors, the materials are to be submitted to the inspection of the engineer at the works before crushing; and all rejected materials are at once to be removed, so that they cannot be used by mistake. Mortar made from lime and underburnt "surkhi" oftentimes sets fairly well, but is liable to disintegrate subsequently from exposure to the weather—a result that has not infrequently occurred in buildings in India.

An engineer may find himself obliged to select a clay, which he will then have to burn to incipient vitrifaction, and after burning, to ensure a careful sorting of the properly burned from the underburned pieces; the latter can be reburned till the whole stock is of approved quality. The clay should be of a pure plastic nature. If only sandy clays are procurable, it may be advisable to wash them—that is, reduce the raw clay to a very finely divided liquid mud, and run the liquid into shallow settling tanks, where the sandy particles will settle to the bottom, and leave the clay-mud comparatively free; this mud is then air-dried, and moulded into thin small plates or tiles for the burning. The plates may be about 4 inches square, and about $\frac{3}{4}$ inch thick.

As a summary, powdered "surkhi" must be from an unctuous clay, must be thoroughly burned, and must be ground or crushed to a powder which will at least pass through an 80-gauge sieve; the finer the powder, the better will be the result of the admixture with the lime.

To ensure good quality in the kankar lime, it should be cleaned and burnt under intelligent and responsible supervision, each stage of the process being subject to careful inspection. "Surkhi" also must be prepared under similar conditions, and the selection of a well-burnt clay of suitable quality is of great importance.

The subject of the addition to pure limes of substances containing soluble silica is dealt with in Art. 9 *et seq.*, also in Arts. 30, 31.

Kankar Lime used under Water.

If kankar lime mortar be submerged in water within 48 hours after mixing, it usually does not set and harden. It must therefore, as a general rule, be exposed to air for at least 48 hours before being covered with water, and it will then continue to harden satisfactorily. The property of hydraulicity will be enhanced by the admixture of powdered well-burnt clay, containing from 60 to 70 per cent. of soluble silica, with the powdered caustic lime; and the more complete the admixture, the better will be the hydraulic quality of the resultant mortar.

KANKAR LIME CONCRETE.

For a kankar lime concrete intended to be impervious to water, the volume of interstices, between the pieces of broken stone, &c., to be used, must be ascertained (see Art. 132). This volume of spaces may be reduced by using stones of graduated sizes, say from $\frac{1}{4}$ inch to $1\frac{1}{2}$ inch cubical pieces well mixed together in equal proportions of fine, medium, and coarse. If the stones be all of 1-inch to $1\frac{1}{2}$-inch cubes, the volume of the spaces may be about 45 per cent. of the mass, and a concrete may be made of

100 cubic feet of stones,
22·5 ,, ,, lime,
22·5 ,, ,, surkhi,

the volume of the mixed mortar to be not less than the volume of the interstices.

The lime and surkhi (or sand) are spread out in successive thin layers on a clean, hard, impervious mixing floor, and are first mixed in a dry state, then water is added through a finely perforated rose jet, and the mixing of the moist materials continued till there is a homogeneous pulpy paste.

When the mortar is thoroughly mixed it is to be

placed in the required proportion on a 6-inch layer of the broken stones, &c., and the mass must be thoroughly incorporated by turning over with shovels, and stirring with rakes having long tines. At least two or three turnings over are required.

The mixed concrete is to be deposited in place, and not thrown down from a height, nor slid down an inclined plane; it is to be spread in layers of about 4 inches uniform thickness, and each layer must be compressed by ramming with heavy wooden rammers till it loses about one-fourth of its deposited thickness. If it be necessary to expose this kankar concrete to running water within forty-eight hours after deposition, it may be given a surface coating of mutton fat mixed with a little lime, as a temporary protection from the solvent action of water. In many cases of constructive works an open textured concrete will be as serviceable as a dense concrete, and the proportion of the broken stone, &c., may be from three to six times the bulk of the lime and surkhi mortar.

In all cases the ingredients must be clean. The stone is to be granular or crystalline, and of durable character; and if the material used be porous, as broken brick, &c., generally is, it must be well soaked in clean water before being mixed into concrete.

II.

TEST TO ASCERTAIN THE AMOUNT OF LIME IN A LIMESTONE, ETC.

SCHEIBLER'S method of ascertaining the proportion of carbonate of lime in the mixture of chalk and clay mud may be used for finding the proportion of lime in kankar and other limes, provided that no other carbonate be present.

The apparatus required consists of a glass tube, about 25 inches long, bent to a U shape (one limb is graduated with division lines to show capacity); a large-mouthed glass jar, about 6 inches diameter and 6 or 7 inches high; and a small cup- or saucer-shaped vessel, that can be lowered into the jar. The jar is stoppered with a perforated cork carrying a glass tube, which is connected with the graduated limb of the U-tube by indiarubber tubing. At a low point on the non-graduated limb of the U-tube is a small outlet pipe, with a tap or clamp. The tube may conveniently be of 150 cubic centimetres capacity in the graduated limb, divided into 0·5 cubic centimetres. It is attached to a board standing vertically on a wooden base-plate.

A small quantity of dried, finely-powdered cement mud, or of limestone, is placed in the cup after being accurately weighed, and about 10 cubic centimetres of hydrochloric acid of 1·12 sp. gr. are placed in the bottle. The cup is lowered into the jar so that powder and acid do not come in contact, the bottle is then stoppered, the stopper being well tallowed so as to be gas-tight, and connection is made to the U-tube, which is filled with distilled water up to the highest graduation mark.

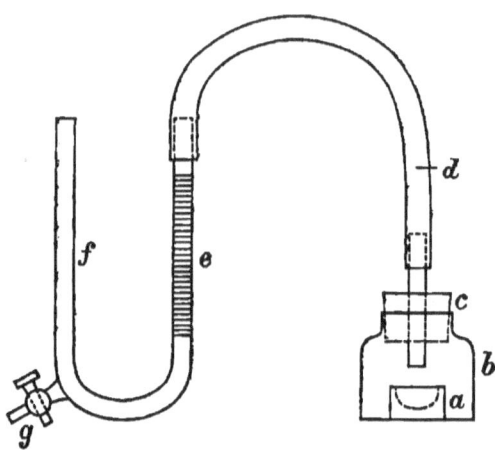

SIMPLE APPARATUS FOR SCHEIBLER'S CALCIMETER TEST.

a, cup or saucer for powder; b, glass jar; c, gas-tight stopper to jar; d, indiarubber tubing; e, graduated limb; f, non-graduated limb; g, tap for discharge of water from f.

The jar is then shaken, and the acid acts upon the cement mud, causing decomposition with the liberation of carbon dioxide (carbonic acid), which passes off into the U-tube. As the gas accumulates, a little water is let off from time to time by the tap, so as to keep equal height of water in both arms. When effervescence ceases, the volume of gas given off is read on the graduated limb by the difference between the original and final height of water.

CARBON DIOXIDE. 201

The apparatus should be placed out of direct sunlight, and be protected from the heat of the observer's body by a glass screen. The temperature at the time of ending the experiment must be noted; a thermometer can be attached to the board supporting the U-tube; as the time occupied is very short, there will probably be no change during the liberation of gas. It is necessary that the jar containing the acid and limestone should be held by the neck, and not grasped bodily by the hand when shaken to mix its contents. Corrections have also to be made for atmospheric pressure; the height of the barometer at the time of the experiment must be noted. A small amount of carbonic acid remains dissolved in the hydrochloric acid; a ratio fixed by Warington is 7 per cent. of the gas measured; this is to be added. Scheibler directs the addition of 8 cubic centimetres to the volume of gas read off.

The formula for ascertaining the corrected weight of carbon dioxide, from the reading on the graduated limb of the volume of carbon dioxide, is

$$\text{Weight of CO}_2 \text{ (in grammes)} = \frac{V \times 0.5381 \times (B - w)}{760 \times (273 + t)},$$

where

V = volume of gas in cubic centimetres.
B = height of barometer in millimetres of mercury.
w = pressure of aqueous vapour at t degrees Cent. (See Table, p. 203.)
t = temperature in degrees Centigrade.

The weight of lime (oxide of calcium) is found by multiplying the weight of the CO_2 by 56 and dividing by 44, or multiplying by 1·27.

A complete description of a somewhat more elaborate apparatus, furnishing more accurate results, is given in Sutton's 'Volumetric Analysis,' p. 98.

Table giving Pressure of Aqueous Vapour in Millimetres of Mercury.

Degrees Cent.	Millimetres.	Degrees Cent.	Millimetres.
0·0	4·60	18·0	15·36
0·5	4·77	18·5	15·85
1·0	4·94	19·0	16·35
1·5	5·12	19·5	16·86
2·0	5·30	20·0	17·39
2·5	5·49	20·5	17·94
3·0	5·69	21·0	18·50
3·5	5·89	21·5	19·07
4·0	6·10	22·0	19·66
4·5	6·31	22·5	20·27
5·0	6·53	23·0	20·89
5·5	6·76	23·5	21·53
6·0	7·00	24·0	22·18
6·5	7·24	24·5	22·86
7·0	7·49	25·0	23·55
7·5	7·75	25·5	24·26
8·0	8·02	26·0	24·99
8·5	8·29	26·5	25·74
9·0	8·57	27·0	26·51
9·5	8·87	27·5	27·29
10·0	9·17	28·0	28·10
10·5	9·47	28·5	28·93
11·0	9·97	29·0	29·78
11·5	10·12	29·5	30·65
12·0	10·46	30·0	31·55
12·5	10·80	30·5	32·46
13·0	11·16	31·0	33·41
13·5	11·53	31·5	34·37
14·0	11·91	32·0	35·36
14·5	12·30	32·5	36·37
15·0	12·70	33·0	37·41
15·5	13·11	33·5	38·47
16·0	13·54	34·0	39·57
16·5	13·97	34·5	40·68
17·0	14·42	35·0	41·83
17·5	14·88		

INDEX.

The figures refer to Nos. of paragraphs, except where otherwise indicated.

ABERDEEN breakwater, mixers at, 153
Aberthaw lime, 18, 22, 157
Abstraction of water from concrete, 143
Acceptance of Portland cement, 89a
Acid lining of kilns, 44
Aden, volcanic sand of, 10, 31
Adhesion test of Portland cement, 84, 85 ; of mortar, 121, 131
Adulteration of Portland cement, 89
Aëration of Portland cement, 45, 66, 68
Age, effect of, on Portland cement, 148
Air flues, kankar kiln, p. 191
Air spaces in concrete exposed to waves, 191
Alkalies, action of, 13
Alumina, influence of, 17, 52, 53, 59, 60, 95
American tests of Portland cement, 83
Ammoniacal liquor, tank for, 166
Analysis, chemical, of Portland cement, 74, 95
Analyses—Aberthaw limestone, 18 ; cement (Portland), 33, 35, 44, 58, 63, 163; ditto briquette, 159 ; chalk, 35 ; Chaux de Teil, 27 ; ciment Grapier, 27 ; clay, 35 ; flints, 59 ; Grapier, ciment, 27 ; gypsum, 23 ; kankar, 30 ; lias limestone, 18 ; limestones, 3, 22 ; pozzolana, 9 ; Rowley rag, 61 ; Santorin earth, 11 ; sea water, 161 ; septaria, 28 ; slag, 26 ; slag cement, 26 ; slate, 59, 61 ; slurry, 35 ; toadstone, 61 ; trass, 10
Andernach, trass, 10
Archimedean screw mortar mixer, 120
Avonmouth Docks, lias lime concrete, 157 ; lime used, 18

BAGS, cement in, 46, 66, 141
Ballast, river and pit, 136
Baker, Sir B., transverse strength of concrete, 172
Baker, Prof., on Portland cement tests, sieves, &c., 79
Barrow lime, 18
Barry, J. Wolfe, lias lime concrete, 157
Barry Docks, concrete, &c., at, 157
Bauschinger, on ratio of tension to compression, 91
Bernays' facing concrete, 147
Berthier's test of hydraulicity, 14
Béton Coignet, 64, 156
Bevan, Mr., on cement powder for briquettes, 68
Blast furnace slag cement, 26; concrete, 26
Block moulding, 165
Block setting, 189
Blue lias limestone, &c. See Lias limestone, &c.
Board, mixing, for concrete, &c., 140
Böhme's hammer test, 71
Boiling water bath, 75
Brick, powdered, 12, 30, 31, and pp. 194–6
Briquettes, shape, moulding and testing, 68–71
Bulwell lime, 19
Burham lime, 16, 129
Burnell, on concrete in sea-water, 170; on lias lime concrete, 157
Burning kankar limestone, p. 193
Burning limestones, 16, 181
Burr millstones, 37, 45, 47

CALCINATION of limestone, 3, 25; of cement mud, 39; phenomena of, 44
Calcined clinker, appearance of, 49; selection of, 45
Carboniferous limestone, 21
Carey, Mr., on temperature of ground cement, 47; on proportions of Portland cement, 52; on the grinding of Portland cement, 77
Carey and Latham's concrete mixer, 151
Casks for Portland cement, 46
Catania, mortar and concrete at, 10a
Cement bags and casks, 46
Cement, definition, 1; Keene's, Martin's, Parian, 29; Roman, 28

INDEX.

Cement, blast furnace slag, 26
Cement, Portland, 33; aëration of, 66; composition, 33, 74, 163; contraction test, 88; ingredients of, 33, 52; German, 71; hot-water bath, 75, 76, 87; manufacture of, 36-39; overclayed, 53; overlimed, 54; reburning of damaged, 99; specification for, 68; storage of, 66, 89a; temperature of, from grinding mills, 48; tests for, 65-98; uses of, 64; weight of, 83
See also "Concrete."
Cement mortar, strength of, 130; adhesion of, 131
Cement and lime mixed mortar, 129
Chalk, white and gray, 16, 35; marl, 16; in France, 27
Chatham Dockyard, concrete at, 147
Chaux de Teil, 27, 59
Chemical analysis of Portland cement, 74, 95. See also "Analyses"
Christiania, mortar mixing in frost, 126
Ciment Grapier, 27
Clay for Portland cement, 33-35; mixed with lime, 7, 8, 14, 16-18
Clay, London, nodules in, 28, 33
Clay, powdered, burnt, 12, 30, 157, and p. 195
Clay, washed, for burning, p. 126; for cement, 43
Clinker, calcined, appearance of, 49; selection of, 45
Clunch lime, 16
Coignet's béton, 156
Coignet's concrete mixer, 155
Colloidal theory of setting of Portland cement, 63
Composition of Portland cement, 33, 74, 163
Concrete, Aberthaw lime, 157; arches, 184, 185; attacks of sea-water on, 157, 158-162; ballast for, 136; in bags, 188; in blocks, 142, 189; broken brick, &c., 137; bridges of, 146, 184; buildings of, 144, 185; cavities in, 191; coal cinders, 138; coke, 138; contraction of, 178, 182; cylinder (hollow), 190; dense, 132; deposition of, 165, 171, 174-176; deposition in sea water, 165, 166, 186; deposition in bags and blocks, 186-188; effect of age on, 148; effect of sea water on, 157-162; facing of, 180, 181; failure of, 158-160, 170; flooring, 181, 183; footways, 183; gap in layers of, 177; honeycombed, 133; for house walls, 144, 184; ingredients of, 134, 142; joining new to old, 177; kankar lime, 30, 196; lias lime, 157; measuring-boxes, 141; mixing by hand, 139, 140; mixing-gang, 141; mixing by machine, 149-155; mixing platform, 140; monolithic, 189; at North Sea Canal Harbour,

133; open-textured, 133; ordinary, in sea water, 142, 158, 159, 164; paving, 179, 182, 183; plastic, 167; poor, 145, 147; Portland cement, 132; proportions of, 134, 142, 145–147, 164; reservoir wall, 142; retaining wall, 182; sea water mixed with, 164, 168; shrinkage of wetted materials for, 142; shrinkage of concrete, 178, 182; slag concrete, 138; stones for, 135, 135a; strength of, 185; tipping of, 173; transverse resistance, 172; walling for houses, 144, 184; watertight, 142, 147, 166, 176; watertight in sea water, 160, 164; weight of, 191; wells, 190
Contraction of cement mortar, 88; of concrete, 178, 182
Crushing rollers for Portland cement, 48
Cylinders, hollow, of concrete, 190

DELIVERY of cement powder, 46, 66
Density test, 80, 94; Schumann's, 94; Fresenius, 94; Mann's, 94
Depositing-skip, 186
Deterioration of cement, 94, 130; of kankar lime, p. 193
Deval hot-water bath, 76
Devonian limestone, 22
Dietrich slurry test, 51
Dietsch kiln, 43
Dorking lime, 16
Draper, proportion of water, 142; concrete in sea-water, 161
Dyce-Cay's test, 87
Dyckerhoff, magnesia in Portland cement, 55, 56; ratio of tension to compression, 91

EDDYSTONE Lighthouse, mortar for, 9, 18
Edge-runner mills, 48, 117–119
Eifel district, trass of, 10
Erdmenger, detection of magnesia, 57
Expansion of concrete, 178
Expansion of lias lime mortar, 18, 100, 157
Expansion test of Portland cement, 87, 88

FAIJA, mixer or gauger, 69; shoot for filling measure, 90
Failure of concrete in sea water, 158–160, 170
Farnham silica stone, 58
Flare lime, 16

INDEX.

Floors, concrete for, 183
Footways, concrete for, 183
Frames for walls of concrete, 144
Francis and Sons, Roman cement, 28
French tests for Portland cement, 71, 82 ; sieves, 78
French mortar mixer, 120
Fresenius, adulteration of Portland cement, 89 ; density of Portland cement, 94
Fuel for calcination, 16, 18, 20, 21, 25, 30, 39-41 ; pp. 190-193
Fulwell, magnesian limestone, 19

GAS-HOLDER tank in Portland cement concrete, 166
Gas producers, use of, for calcination, 40, 41
Gauge of wire for sieves, 72, 79
German tests for Portland cement, 82 ; of sand, 72 ; of sieves, 78
Gillmore's needle test, 83
Goreham's process, 37
Grant, mould for briquette, 69 ; details of testing, 71 ; fineness of grinding, 90
Grapier ciment, 27
Griffith, disintegration of Portland cement briquette, 96
Grinding kankar lime, p. 193
Grinding mills for cement, 48
Grinding, test for fineness of, 77, 90
Grouting with mortar, 113
Guildford lime, 16
Gypsum, 23
Gypsum, in Portland cement, 162

HALLING lime, 16
Halkin lime, 21
Harrison, Mr. C., on lias lime concrete, 157
Hartlepool Docks, lias lime at, 157
Hayter Lewis, on cement mortar, 90
Hewitt, A. H., analyses by, &c., 44
Hot-water bath for briquettes, &c., 75, 76, 87
House walls in concrete, 144, 184, 185
Hutton, Darnton, on sand in concrete, 133
Hydraulic lime, 8
Hydraulicity, property of, 9 ; test for, 14

INDIAN limes, 30-32, and pp. 189-198
Ingredients, proportion of, in Portland cement, 33, 52-54

JAPP'S volumetric process, 51
Johnson, Mr., on sieve test, 77
Joints, open, in concrete exposed to waves, 191
Jones, H. E., lining to concrete tank, 166
Joy's kiln, 42

KANKAR lime, 30, and pp. 189-198; concrete, pp. 196, 197; kilns, pp. 190, 191, and diagram
Keene's cement, 29
Keynsham lime, 18
Kilns—Dietsch's kiln, 43; draw, 21; flare, 16; Joy's, 42; kankar, pp. 190, 191, and diagram; Ransome's, 40; running, 25; shapes of, 25; Stokes', 41
Kimmeridge clay, 17
Kinipple, on plastic concrete, 167
Knapp, alumina, &c., in Portland cement, 95
Kunkur, *see* Kankar

LAITANCE, on surface of concrete, 177
Larrying mortar, 113
Lee & Co.'s concrete mixer, 152
Lias lime concrete, 157
Lias lime mortar, 100, 101, 115, 116, 157; expansion of, 18, 100, 157
Lias limestone, 18; lime, 18
Light concrete, 138, 191
Lime, action of, on sand, 109
Lime and cement mortar, 129
Lime, clayey, 7, 8; concrete, 30; definition, 1; flare, 16; hardening of, 6; in India, 30, and pp. 189-198; sale of, 15; slaking of, 3, 4, 7, 8, 17, 100
Limestone, calcination of, 3, 16, 18, 21, 25
Limestones, carboniferous, 21; Devonian, 22; lias, 18; magnesian, 19; oölitic, 17; Silurian, 22; in India, 30-32, and p. 189
Liverpool Dock walls, lime for, 21
Lunge's nitrometer process, 51
Lyme Regis lias limestone, 18

INDEX.

MACHINES for mixing (Faija's), 69. Also 117, 120, 149-155
Magnesia, detection of, 57; in cement, 55; in lime, 19; safe proportion, 56
Magnesian limestone, 19, 20
Magnesian sulphates and chlorides in Portland cement, 161
Mann, adhesion test, 85; density test, 94; needle test, 96
Marcus Vitruvius, fallacy of, 16
Margetts, on hot-water bath, 75, 76
Marsden, magnesian limestone, 19
Martin's cement, 29
Matthews, Dr., on alkalies in Portland cement, 13
McLeod, Prof., experiments on Portland cement, 63
Medina cement, 28
Medway lime, 16
Memoranda of weights, &c., p. 187
Merstham lime, 16
Messent, Mr., mixer, 153; proportion of sand, 160
Michaelis, Dr., on clay, 35; on influence of silica, 59; on magnesia, 55; on setting of cement, 63; on sieve test, 77; on tests of Portland cement, 91; on watertight mortar, 125; on weight of Portland cement, 90; remedial measures, 169
Microscope, mortar under the, 90
Mills, edge-runner, 117, 48; roller, 48; millstones, 47; pug-mill, 155
Mixers, Portland cement concrete—Carey and Latham's, 151; Coignet's, 155; Lee & Co.'s, 152; Messent's, 153; Stoney's, 150; sundry, 154
Mixing concrete, 139; board for, 140; by hand, 140
Mixing mortar, by hand, 116; by machine, 117, 118; with pozzolana, 119
Molasses, addition of, to mortar, 32, 123
Mortar, adhesion, 131; briquettes, 67-69, 83; cement, 122; grouting, 113; larrying, 113; lias lime, 115, 157; lime, 100; mixed lime and cement, 129; mixing in frost, 126; ditto by hand, 116; ditto by machine, 117, 118, 120; ditto with pozzolana, 119, 157; ditto with Aden pumice, 31; with molasses, 32, 123; plastering and pointing, 127, 128; proportion of cement, 122-125; ditto of lime, 101, 111; ditto of sand, 101, 111; pure lime, re-use of, 114; in sea water, 125, 157; strength of, 121, 130; water for, 112; watertight, 125, 166
Mould for briquette, 69, 83

NEEDLE test, 83; German, 98; Mann's, 96; Vicat's, 97
Neate, Mr. P., on temperature of roller-ground cement, 48
Newhaven Harbour, concrete works, 151
Nodules of clayey limestone, 17, 28, 33

Oölitic limes, 17
Overclayed Portland cement, 53, 60, 95
Overlimed Portland cement, 54

PAN mortar mills, 117-119
Parian cement, 29
Pat tests, 87, 88
Paving, cement concrete for, 179, 181-183
Peterborough, lime at, 17
Pier, of slag cement concrete, 26
Plaster of Paris, 23
Plastering, cement mortar, 127, 128; with Roman cement, 28
Plastic concrete, 167
Platform, mixing, 140
Pockets, concrete for, 145
Pointing of joints, 128
Portland cement, 33-63; composition, 35, 52-54; ditto, theoretical, 163; concrete, 132-155; definition of, 33; localities of manufacture, 34; mixture of chalk and clay for, 35-37; mortar, 122-131; proportion of ingredients for, 23, 52; re-burning of damaged, 99; testing of slurry for, 38, 50, 51, 73; testing of briquette of, 70, 71, 75, 76; uses of, 64
Potter, T., on building in concrete, 144
Pozzolana, 9, 119, 157
Prussian test for contraction, &c., 88
Pure lime, 3, 4, 6, 9, 12; mortar, re-use of, 114

QUICK-SETTING cement, 28, 96

RANSOME's calciner, 40, 51
Ransome's patent stone, 58

INDEX. 213

Re-burning damaged cement, 99
Redgrave, Mr. G. R., slag cement, 26
Reid, Mr., on cement, 61 ; on cement clinker, 46; on influence of silica, 59 ; on submerged clinker, 46
Remedial measures for decaying concrete, 169
Removal of rejected cement, 89a
Removal of rejected surkhi, p. 195
Retaining walls, in concrete, 182
Re-use of pure lime mortar, 114
Roman cement, 28, 33
Rosendale cement, 20
Rowley ragstone, 61

SALE of lime and sand, 15
Sand, action of lime on, 109 ; blown sand, 105 ; for briquettes, 92 ; clayey, 107, 122, 123 ; from crushed rock, 108, 110 ; for dense concrete, 132 ; granulated slag, 110, 138 ; in India, 30, p. 194; pit-sand, 103, 104 ; proportions of sand in concrete, 26, 132-135a, 142, 147, 156, 160, 183 ; ditto in India, 30, p. 194 ; ditto in mortar, 101, 111, 115, 124, 125, 130, 166 ; quality of, 102 ; sale of, 15 ; sea-sand, 106, 164, 168 ; shrinkage of wetted sand, 142 ; slag sand, granulated, 110, 138; standard sand, 72, 92 ; use of, 101 ; virgin sand, 102 ; volume of sand for dense concrete, 142, 156, 160 ; for watertight concrete and mortar, 125, 142, 160
Sandeman, Mr., on watertight mortar, 125
Santorin earth, 11, 63
Schelbler's method of testing slurry, &c., 51, also p. 199
Schumann's test for density, 94
Scum on surface of concrete, 177
Sea water, composition of, 161
Sea water, concrete for, 142, 156, 158-160 ; use of, 164, 168 ; see also sea sand, 106 ; precautions for concrete in, 164
Selenitic lime, &c., 24
Septaria nodules, 17, 28, 33
Setting of cement, 28, 62, 63; ditto of lime, 6-9 ; test for setting, 81, 83 ; time of, for cement, 96
Shed, storage, 45, 66, 89a
Shortness of cement mortar, 122
Shrinkage of concrete, 178, 182 ; of wetted sand, &c., 142

Sieve tests, 77–79, 90; wire for, 79
Silica, influence of, 17, 52, 58–61, 95; soluble form, 13; stone (Farnham), 58
Silurian limestone, 22
Skinningrove, slag cement at, 26
Skip, depositing, 186
Slag cement, 26; concrete, 26; granulated, 138
Slaking, of pure lime, 3, 4, 7, 8, 16, 100; lias lime, 17, 100
Slow setting of cement, 96; time of, 96
Slurry, testing of, 38, 50, 51, 73, p. 198
Smeaton, 9, 16, 18, 23
Soda crystals, use of, 126
Sorel's cement, 55
Soorkee, *see* Surkhi
Specific gravity test, 80, 94
Speeton clay nodules, 17
Stanger and Blount, Messrs., letter of, 44
Stern brand of cement, 71
Stevenson, test for Portland cement, 86
Stokes' calciner, 41; experiments on briquettes, 76; on acid lining of kiln, 44
Stones for concrete, 135, 135*a*; ditto, large, in concrete, 165
Stone-breakers, 135*a*
Stoney's mixer, 150
Storage of cement, 45, 46, 66, 89*a*
Sugar syrup, addition of, to mortar, 32, 123
Sulphuric acid in Portland cement, 95
Sunderland Docks, lias lime mortar at, 157
Surkhi, 30, p. 194

TABLE of pressure of aqueous vapour, p. 203
Teil, Chaux de, 27
Temperature of cement powder, 47, 48; change of, during setting, 81
Tensile stress-test, 67, 68, 70, 71, 91
Test of slurry, 38, 50, 51, 73, p. 199
Test of a limestone, p. 199
Test-loads for cement briquettes, 68, 70, 71, 76, 82, 83
Tests for Portland cement, 67; adhesion, 84, 85; adulteration, 89; chemical, 74, 95; contraction, 87, 88; density, 80, 94; expansion,

INDEX.

87, 88; fineness of grinding, 77–79, 82, 83, 90; water bath, 75, 76, 87, 92; tensile stress, 67–71, 82, 83, 91; transverse bending, 86; weight, 67, 90
Trass, 10
Treussart, Gen., hydraulic lime concrete, 157
Trieste harbour works concrete, 27
Tyne Docks, pozzolana in lias lime mortar, &c., 157

UNWIN's formula for the strength of cement briquette, 71
Uses of Portland cement, 64

VICAT, needle test, 97; hydraulic lime concrete, 157; on failure of concrete, 170
Victoria stone, patent, 58
Volcanic ash, 9, 10, 10a, 11, 31
Vyrnwy reservoir, mortar for, 125; sieve test for cement, 79

WALLS, house, in concrete, 144, 184
Wash mill, 37
Water, kankar lime in, p. 195
Water for mixing mortar, 112, 124, 141–143
Water for test briquettes, 69, 93
Watertight coating to concrete, 147, 166
Watertight Portland cement concrete, 142, 147, 160, 164, 166, 176
Watson, Mr., sieve test, 77
Weight of concrete, 191; in water, 191
Weight test of Portland cement, 67, 90
Wells, concrete, 190
Wetting of brickwork for pointing, &c., 112, 127, 128
Wetting of Portland cement concrete, 63, 142, 177
Wire for sieves, size of, 79
Wouldham lime, 16

LONDON: PRINTED BY WILLIAM CLOWES AND SONS, LIMITED,
STAMFORD STREET AND CHARING CROSS.

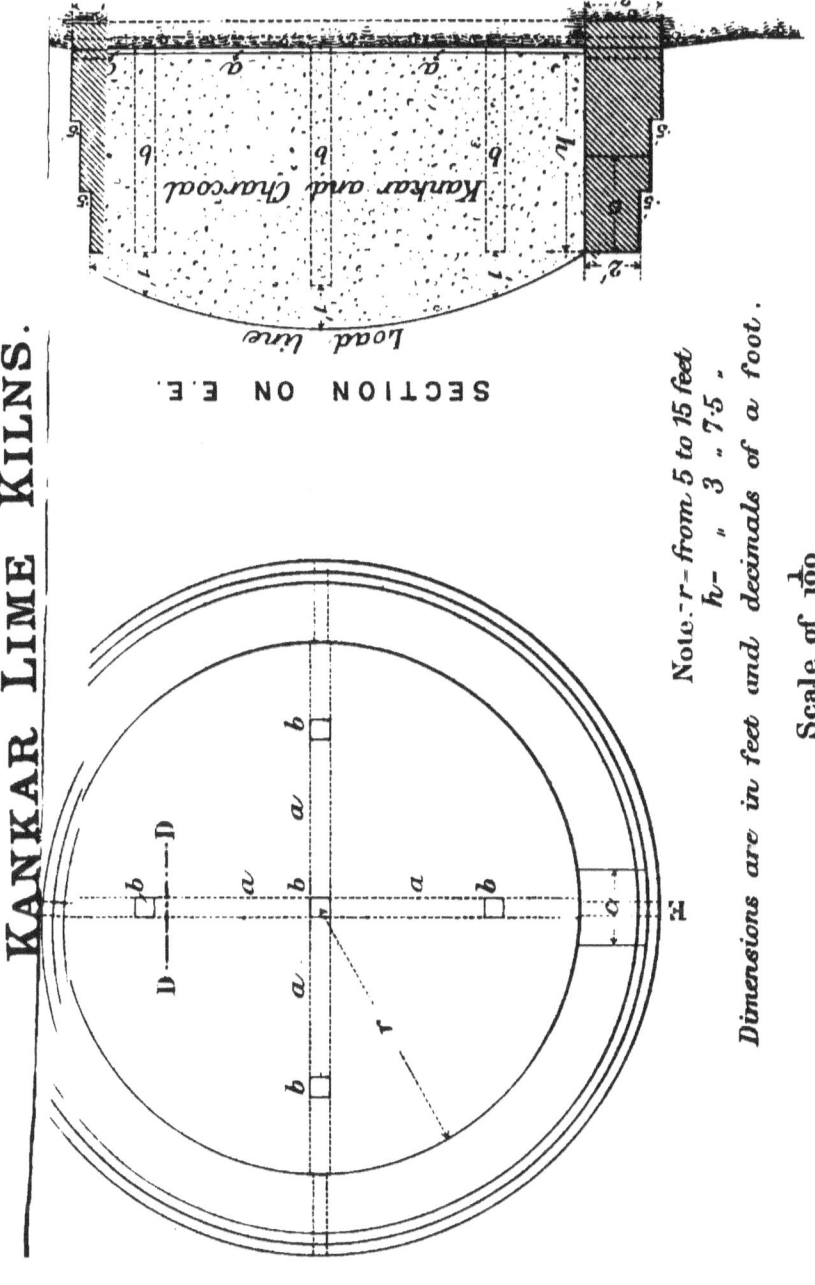

1892.

BOOKS RELATING
TO
APPLIED SCIENCE,
PUBLISHED BY
E. & F. N. SPON,
LONDON: 125, STRAND.

NEW YORK: 12, CORTLANDT STREET.

The Engineers' Sketch-Book of Mechanical Movements, Devices, Appliances, Contrivances, Details employed in the Design and Construction of Machinery for every purpose. Collected from numerous Sources and from Actual Work. Classified and Arranged for Reference. *Nearly* 2000 *Illustrations.* By T. B. BARBER, Engineer. Second Edition, 8vo, cloth, 7s. 6d.

A Pocket-Book for Chemists, Chemical Manufacturers, Metallurgists, Dyers, Distillers, Brewers, Sugar Refiners, Photographers, Students, etc., etc. By THOMAS BAYLEY, Assoc. R.C. Sc. Ireland, Analytical and Consulting Chemist and Assayer. Fifth edition, 481 pp., royal 32mo, roan, gilt edges, 5s.

SYNOPSIS OF CONTENTS:

Atomic Weights and Factors—Useful Data—Chemical Calculations—Rules for Indirect Analysis—Weights and Measures—Thermometers and Barometers—Chemical Physics—Boiling Points, etc.—Solubility of Substances—Methods of Obtaining Specific Gravity—Conversion of Hydrometers—Strength of Solutions by Specific Gravity—Analysis—Gas Analysis—Water Analysis—Qualitative Analysis and Reactions—Volumetric Analysis—Manipulation—Mineralogy — Assaying — Alcohol — Beer — Sugar — Miscellaneous Technological matter relating to Potash, Soda, Sulphuric Acid, Chlorine, Tar Products, Petroleum, Milk, Tallow, Photography, Prices, Wages, Appendix, etc., etc.

Electricity, its Theory, Sources, and Applications.
By JOHN T. SPRAGUE, M. Inst. E.E. Third edition, thoroughly revised and extended, *with numerous illustrations and tables*, crown 8vo, cloth, 15s.

Electric Toys. Electric Toy-Making, Dynamo Building and Electric Motor Construction for Amateurs. By T. O'CONOR SLOANE, Ph.D. *With cuts*, square 16mo, cloth, 4s. 6d.

The Phonograph, and How to Construct it. With a Chapter on Sound. By W. GILLETT. *With engravings and full working drawings*, crown 8vo, cloth, 5s.

B

CATALOGUE OF SCIENTIFIC BOOKS

Just Published, in Demy 8vo, cloth, containing 975 pages and 250 Illustrations, price 7s. 6d.

SPONS' HOUSEHOLD MANUAL:
A Treasury of Domestic Receipts and Guide for Home Management.

PRINCIPAL CONTENTS.

Hints for selecting a good House, pointing out the essential requirements for a good house as to the Site, Soil, Trees, Aspect, Construction, and General Arrangement; with instructions for Reducing Echoes, Waterproofing Damp Walls, Curing Damp Cellars.

Sanitation.—What should constitute a good Sanitary Arrangement; Examples (with Illustrations) of Well- and Ill-drained Houses; How to Test Drains; Ventilating Pipes, etc.

Water Supply.—Care of Cisterns; Sources of Supply; Pipes; Pumps; Purification and Filtration of Water.

Ventilation and Warming.—Methods of Ventilating without causing cold draughts, by various means; Principles of Warming; Health Questions; Combustion; Open Grates; Open Stoves; Fuel Economisers; Varieties of Grates; Close-Fire Stoves; Hot-air Furnaces; Gas Heating; Oil Stoves; Steam Heating; Chemical Heaters; Management o Flues; and Cure of Smoky Chimneys.

Lighting.—The best methods of Lighting; Candles, Oil Lamps, Gas, Incandescent Gas, Electric Light; How to test Gas Pipes; Management of Gas.

Furniture and Decoration.—Hints on the Selection of Furniture; on the most approved methods of Modern Decoration; on the best methods of arranging Bells and Calls How to Construct an Electric Bell.

Thieves and Fire.—Precautions against Thieves and Fire; Methods of Detection Domestic Fire Escapes; Fireproofing Clothes, etc.

The Larder.—Keeping Food fresh for a limited time; Storing Food without change, such as Fruits, Vegetables, Eggs, Honey, etc.

Curing Foods for lengthened Preservation, as Smoking, Salting, Canning, Potting, Pickling, Bottling Fruits, etc.; Jams, Jellies, Marmalade, etc.

The Dairy.—The Building and Fitting of Dairies in the most approved modern style; Butter-making; Cheesemaking and Curing.

The Cellar.—Building and Fitting; Cleaning Casks and Bottles; Corks and Corking; Aërated Drinks; Syrups for Drinks; Beers; Bitters; Cordials and Liqueurs; Wines; Miscellaneous Drinks.

The Pantry.—Bread-making; Ovens and Pyrometers; Yeast; German Yeast; Biscuits; Cakes; Fancy Breads; Buns.

The Kitchen.—On Fitting Kitchens; a description of the best Cooking Ranges, close and open; the Management and Care of Hot Plates, Baking Ovens, Dampers, Flues, and Chimneys; Cooking by Gas; Cooking by Oil; the Arts of Roasting, Grilling, Boiling, Stewing, Braising, Frying.

Receipts for Dishes—Soups, Fish, Meat, Game, Poultry, Vegetables, Salads, Puddings, Pastry, Confectionery, Ices, etc., etc.; Foreign Dishes.

The Housewife's Room.—Testing Air, Water, and Foods; Cleaning and Renovating; Destroying Vermin.

Housekeeping, Marketing.

The Dining-Room.—Dietetics; Laying and Waiting at Table: Carving; Dinners, Breakfasts, Luncheons, Teas, Suppers, etc.

The Drawing-Room.—Etiquette; Dancing; Amateur Theatricals; Tricks and Illusions; Games (indoor).

The Bedroom and Dressing-Room; Sleep; the Toilet; Dress; Buying Clothes; Outfits; Fancy Dress.

The Nursery.—The Room; Clothing; Washing; Exercise; Sleep; Feeding; Teething; Illness; Home Training.

The Sick-Room.—The Room; the Nurse; the Bed; Sick Room Accessories; Feeding Patients; Invalid Dishes and Drinks; Administering Physic; Domestic Remedies; Accidents and Emergencies; Bandaging; Burns; Carrying Injured Persons; Wounds; Drowning; Fits Frost-bites; Poisons and Antidotes; Sunstroke; Common Complaints; Disinfection, etc.

The Bath-Room.—Bathing in General; Management of Hot-Water System.
The Laundry.—Small Domestic Washing Machines, and methods of getting up linen Fitting up and Working a Steam Laundry.
The School-Room.—The Room and its Fittings; Teaching, etc.
The Playground.—Air and Exercise; Training; Outdoor Games and Sports.
The Workroom.—Darning, Patching, and Mending Garments.
The Library.—Care of Books.
The Garden.—Calendar of Operations for Lawn, Flower Garden, and Kitchen Garden.
The Farmyard.—Management of the Horse, Cow, Pig, Poultry, Bees, etc., etc.
Small Motors.—A description of the various small Engines useful for domestic purposes, from 1 man to 1 horse power, worked by various methods, such as Electric Engines, Gas Engines, Petroleum Engines, Steam Engines, Condensing Engines, Water Power, Wind Power, and the various methods of working and managing them.
Household Law.—The Law relating to Landlords and Tenants, Lodgers, Servants, Parochial Authorities, Juries, Insurance, Nuisance, etc.

On Designing Belt Gearing. By E. J. Cowling
Welch, Mem. Inst. Mech. Engineers, Author of 'Designing Valve Gearing.' Fcap. 8vo, sewed, 6*d.*

A Handbook of Formulæ, Tables, and Memoranda,
for Architectural Surveyors and others engaged in Building. By J. T. Hurst, C.E. Fourteenth edition, royal 32mo, roan, 5*s.*

"It is no disparagement to the many excellent publications we refer to, to say that in our opinion this little pocket-book of Hurst's is the very best of them all, without any exception. It would be useless to attempt a recapitulation of the contents, for it appears to contain almost *everything* that anyone connected with building could require, and, best of all, made up in a compact form for carrying in the pocket, measuring only 5 in. by 3 in., and about ¾ in. thick, in a limp cover. We congratulate the author on the success of his laborious and practically compiled little book, which has received unqualified and deserved praise from every professional person to whom we have shown it."—*The Dublin Builder.*

Tabulated Weights of Angle, Tee, Bulb, Round,
Square, and Flat Iron and Steel, and other information for the use of Naval Architects and Shipbuilders. By C. H. Jordan, M.I.N.A. Fourth edition, 32mo, cloth, 2*s.* 6*d.*

A Complete Set of Contract Documents for a Country
Lodge, comprising Drawings, Specifications, Dimensions (for quantities), Abstracts, Bill of Quantities, Form of Tender and Contract, with Notes by J. Leaning, printed in facsimile of the original documents, on single sheets fcap., in paper case, 10*s.*

A Practical Treatise on Heat, as applied to the
Useful Arts; for the Use of Engineers, Architects, &c. By Thomas Box. *With 14 plates.* Sixth edition, crown 8vo, cloth, 12*s.* 6*d.*

A Descriptive Treatise on Mathematical Draw-
Instruments: their construction, uses, qualities, selection, pre and suggestions for improvements, with hints upon Drawing ing. By W. F. Stanley, M.R.I. Sixth edition, *with numer-* crown 8vo, cloth, 5*s.*

Quantity Surveying. By J. LEANING. With 42 illustrations. Second edition, revised, crown 8vo, cloth, 9s.

CONTENTS:

A complete Explanation of the London Practice.
General Instructions.
Order of Taking Off.
Modes of Measurement of the various Trades.
Use and Waste.
Ventilation and Warming.
Credits, with various Examples of Treatment.
Abbreviations.
Squaring the Dimensions.
Abstracting, with Examples in illustration of each Trade.
Billing.
Examples of Preambles to each Trade.
Form for a Bill of Quantities.
   Do.   Bill of Credits.
   Do.   Bill for Alternative Estimate.
Restorations and Repairs, and Form of Bill.
Variations before Acceptance of Tender.
Errors in a Builder's Estimate.
Schedule of Prices.
Form of Schedule of Prices.
Analysis of Schedule of Prices.
Adjustment of Accounts.
Form of a Bill of Variations.
Remarks on Specifications.
Prices and Valuation of Work, with Examples and Remarks upon each Trade.
The Law as it affects Quantity Surveyors, with Law Reports.
Taking Off after the Old Method.
Northern Practice.
The General Statement of the Methods recommended by the Manchester Society of Architects for taking Quantities.
Examples of Collections.
Examples of "Taking Off" in each Trade.
Remarks on the Past and Present Methods of Estimating.

Spons' Architects' and Builders' Price Book, with *useful Memoranda.* Edited by W. YOUNG, Architect. Crown 8vo, cloth, red edges, 3s. 6d. *Published annually.* Nineteenth edition. *Now ready.*

Long-Span Railway Bridges, comprising Investigations of the Comparative Theoretical and Practical Advantages of the various adopted or proposed Type Systems of Construction, with numerous Formulæ and Tables giving the weight of Iron or Steel required in Bridges from 300 feet to the limiting Spans; to which are added similar Investigations and Tables relating to Short-span Railway Bridges. Second and revised edition. By B. BAKER, Assoc. Inst. C.E. *Plates,* crown 8vo, cloth, 5s.

Elementary Theory and Calculation of Iron Bridges and Roofs. By AUGUST RITTER, Ph.D., Professor at the Polytechnic School at Aix-la-Chapelle. Translated from the third German edition, by H. R. SANKEY, Capt. R.E. With 500 *illustrations,* 8vo, cloth, 15s.

The Elementary Principles of Carpentry. By THOMAS TREDGOLD. Revised from the original edition, and partly re-written, by JOHN THOMAS HURST. Contained in 517 pages of letter-press, and *illustrated with* 48 *plates and* 150 *wood engravings.* Sixth edition, reprinted from the third, crown 8vo, cloth, 12s. 6d.

Section I. On the Equality and Distribution of Forces—Section II. Resistance of Timber—Section III. Construction of Floors—Section IV. Construction of Roofs—Section V. Construction of Domes and Cupolas—Section VI. Construction of Partitions—Section VII. Scaffolds, Staging, and Gantries—Section VIII. Construction of Centres for Bridges—Section IX. Coffer-dams, Shoring, and Strutting—Section X. Wooden Bridges and Viaducts—Section XI. Joints, Straps, and other Fastenings—Section XII. Timber.

The Builder's Clerk: a Guide to the Management of a Builder's Business. By THOMAS BALES. Fcap. 8vo, cloth, 1s. 6d.

Practical Gold-Mining: a Comprehensive Treatise
on the Origin and Occurrence of Gold-bearing Gravels, Rocks and Ores.
and the methods by which the Gold is extracted. By C. G. WARNFORD
LOCK, co-Author of 'Gold: its Occurrence and Extraction.' *With 8 plates
and 275 engravings in the text*, royal 8vo, cloth, 2*l.* 2*s.*

Hot Water Supply: A Practical Treatise upon the
Fitting of Circulating Apparatus in connection with Kitchen Range and
other Boilers, to supply Hot Water for Domestic and General Purposes.
With a Chapter upon Estimating. *Fully illustrated*, crown 8vo, cloth, 3*s.*

Hot Water Apparatus: An Elementary Guide for
the Fitting and Fixing of Boilers and Apparatus for the Circulation of
Hot Water for Heating and for Domestic Supply, and containing a
Chapter upon Boilers and Fittings for Steam Cooking. 32 *illustrations*,
fcap. 8vo, cloth, 1*s.* 6*d.*

The Use and Misuse, and the Proper and Improper
Fixing of a Cooking Range. Illustrated, fcap. 8vo, sewed, 6*d.*

Iron Roofs: Examples of Design, Description. *Illustrated with* 64 *Working Drawings of Executed Roofs.* By ARTHUR T.
WALMISLEY, Assoc. Mem. Inst. C.E. Second edition, revised, imp. 4to,
half-morocco, 3*l.* 3*s.*

A History of Electric Telegraphy, to the Year 1837.
Chiefly compiled from Original Sources, and hitherto Unpublished Documents, by J. J. FAHIE, Mem. Soc. of Tel. Engineers, and of the International Society of Electricians, Paris. Crown 8vo, cloth, 9*s.*

Spons' Information for Colonial Engineers. Edited
by J. T. HURST. Demy 8vo, sewed.

No. 1, Ceylon. By ABRAHAM DEANE, C.E. 2*s.* 6*d.*

CONTENTS:
Introductory Remarks — Natural Productions — Architecture and Engineering — Topography, Trade, and Natural History — Principal Stations — Weights and Measures, etc., etc.

No. 2. Southern Africa, including the Cape Colony, Natal, and the Dutch Republics. By HENRY HALL, F.R.G.S., F.R.C.I. With Map. 3*s.* 6*d.*

CONTENTS:
General Description of South Africa — Physical Geography with reference to Engineering Operations — Notes on Labour and Material in Cape Colony — Geological Notes on Rock Formation in South Africa — Engineering Instruments for Use in South Africa — Principal Public Works in Cape Colony: Railways, Mountain Roads and Passes, Harbour Works, Bridges, Gas Works, Irrigation and Water Supply, Lighthouses, Drainage and Sanitary Engineering, Public Buildings, Mines — Table of Woods in South Africa — Animals used for Draught Purposes — Statistical Notes — Table of Distances — Rates of Carriage, etc.

No. 3. India. By F. C. DANVERS, Assoc. Inst. C.E. With Map. 4*s.* 6*d.*

CONTENTS:
Physical Geography of India — Building Materials — Roads — Railways — Bridges — Irrigation — River Works — Harbours — Lighthouse Buildings — Native Labour — The Principal Trees of India — Money — Weights and Measures — Glossary of Indian Terms, etc.

Our Factories, Workshops, and Warehouses: their
Sanitary and Fire-Resisting Arrangements. By B. H. THWAITE, Assoc.
Mem. Inst. C.E. With 183 *wood engravings*, crown 8vo, cloth, 9s.

A Practical Treatise on Coal Mining. By GEORGE
G. ANDRÉ, F.G.S., Assoc. Inst. C.E., Member of the Society of Engineers.
With 82 *lithographic plates*. 2 vols., royal 4to, cloth, 3l. 12s.

A Practical Treatise on Casting and Founding,
including descriptions of the modern machinery employed in the art. By
N. E. SPRETSON, Engineer. Fifth edition, with 82 *plates* drawn to
scale, 412 pp., demy 8vo, cloth, 18s.

A Handbook of Electrical Testing. By H. R. KEMPE,
M.S.T.E. Fourth edition, revised and enlarged, crown 8vo, cloth, 16s.

The Clerk of Works: a Vade-Mecum for all engaged
in the Superintendence of Building Operations. By G. G. HOSKINS,
F.R.I.B.A. Third edition, fcap. 8vo, cloth, 1s. 6d.

American Foundry Practice: Treating of Loam,
Dry Sand, and Green Sand Moulding, and containing a Practical Treatise
upon the Management of Cupolas, and the Melting of Iron. By T. D.
WEST, Practical Iron Moulder and Foundry Foreman. Second edition,
with numerous illustrations, crown 8vo, cloth, 10s. 6d.

The Maintenance of Macadamised Roads. By T.
CODRINGTON, M.I.C.E, F.G.S., General Superintendent of County Roads
for South Wales. Second edition, 8vo, cloth, 7s. 6d.

Hydraulic Steam and Hand Power Lifting and
Pressing Machinery. By FREDERICK COLYER, M. Inst. C.E., M. Inst. M.E.
With 73 *plates*, 8vo, cloth, 18s.

Pumps and Pumping Machinery. By F. COLYER,
M.I.C.E., M.I.M.E. With 23 *folding plates*, 8vo, cloth, 12s. 6d.

Pumps and Pumping Machinery. By F. COLYER.
Second Part. With 11 *large plates*, 8vo, cloth, 12s. 6d.

A Treatise on the Origin, Progress, Prevention, and
Cure of Dry Rot in Timber; with Remarks on the Means of Preserving
Wood from Destruction by Sea-Worms, Beetles, Ants, etc. By THOMAS
ALLEN BRITTON, late Surveyor to the Metropolitan Board of Works,
etc., etc. With 10 *plates*, crown 8vo, cloth, 7s. 6d.

The Artillery of the Future and the New Powders.
By J. A. LONGRIDGE, Mem. Inst. C.E. 8vo, cloth, 5s.

Gas Works: their Arrangement, Construction, Plant,
and Machinery. By F. COLYER, M. Inst. C.E. *With* 31 *folding plates,* 8vo, cloth, 12s. 6d.

The Municipal and Sanitary Engineer's Handbook.
By H. PERCY BOULNOIS, Mem. Inst. C.E., Borough Engineer, Portsmouth. *With numerous illustrations.* Second edition, demy 8vo, cloth, 15s.

CONTENTS:

The Appointment and Duties of the Town Surveyor—Traffic—Macadamised Roadways—Steam Rolling—Road Metal and Breaking—Pitched Pavements—Asphalte—Wood Pavements—Footpaths—Kerbs and Gutters—Street Naming and Numbering—Street Lighting—Sewerage—Ventilation of Sewers—Disposal of Sewage—House Drainage—Disinfection—Gas and Water Companies, etc., Breaking up Streets—Improvement of Private Streets—Borrowing Powers—Artizans' and Labourers' Dwellings—Public Conveniences—Scavenging, including Street Cleansing—Watering and the Removing of Snow—Planting Street Trees—Deposit of Plans—Dangerous Buildings—Hoardings—Obstructions—Improving Street Lines—Cellar Openings—Public Pleasure Grounds—Cemeteries—Mortuaries—Cattle and Ordinary Markets—Public Slaughter-houses, etc.—Giving numerous Forms of Notices, Specifications, and General Information upon these and other subjects of great importance to Municipal Engineers and others engaged in Sanitary Work.

Metrical Tables. By Sir G. L. MOLESWORTH,
M.I.C.E. 32mo, cloth, 1s. 6d.

CONTENTS.

General—Linear Measures—Square Measures—Cubic Measures—Measures of Capacity—Weights—Combinations—Thermometers.

Elements of Construction for Electro-Magnets. By
Count TH. DU MONCEL, Mem. de l'Institut de France. Translated from the French by C. J. WHARTON. Crown 8vo, cloth, 4s. 6d.

A Treatise on the Use of Belting for the Transmission of Power. By J. H. COOPER. Second edition, *illustrated,* 8vo, cloth, 15s.

A Pocket-Book of Useful Formulæ and Memoranda
for Civil and Mechanical Engineers. By Sir GUILFORD L. MOLESWORTH, Mem. Inst. C.E. *With numerous illustrations,* 744 pp. Twenty-second edition, 32mo, roan, 6s.

SYNOPSIS OF CONTENTS:

Surveying, Levelling, etc.—Strength and Weight of Materials—Earthwork, Brickwork, Masonry, Arches, etc.—Struts, Columns, Beams, and Trusses—Flooring, Roofing, and Roof Trusses—Girders, Bridges, etc.—Railways and Roads—Hydraulic Formulæ—Canals, Sewers, Waterworks, Docks—Irrigation and Breakwaters—Gas, Ventilation, and Warming—Heat, Light, Colour, and Sound—Gravity: Centres, Forces, and Powers—Millwork, Teeth of Wheels, Shafting, etc.—Workshop Recipes—Sundry Machinery—Animal Power—Steam and the Steam Engine—Water-power, Water-wheels, Turbines, etc.—Wind and Windmills—Steam Navigation, Ship Building, Tonnage, etc.—Gunnery, Projectiles, etc.—Weights, Measures, and Money—Trigonometry, Conic Sections, and Curves—Telegraphy—Mensuration—Tables of Areas and Circumference, and Arcs of Circles—Logarithms, Square and Cube Roots, Powers—Reciprocals, etc.—Useful Numbers—Differential and Integral Calculus—Algebraic Signs—Telegraphic Construction and Formulæ.

Hints on Architectural Draughtsmanship. By G. W.
TUXFORD HALLATT. Fcap. 8vo, cloth, 1s. 6d.

Spons' Tables and Memoranda for Engineers;
selected and arranged by J. T. HURST, C.E., Author of 'Architectural Surveyors' Handbook,' 'Hurst's Tredgold's Carpentry,' etc. Eleventh edition, 64mo, roan, gilt edges, 1s.; or in cloth case, 1s. 6d.

This work is printed in a pearl type, and is so small, measuring only 2½ in. by 1¾ in. by ¼ in. thick, that it may be easily carried in the waistcoat pocket.

"It is certainly an extremely rare thing for a reviewer to be called upon to notice a volume measuring but 2½ in. by 1¾ in., yet these dimensions faithfully represent the size of the handy little book before us. The volume—which contains 118 printed pages, besides a few blank pages for memoranda—is, in fact, a true pocket-book, adapted for being carried in the waistcoat pocket, and containing a far greater amount and variety of information than most people would imagine could be compressed into so small a space. . . . . The little volume has been compiled with considerable care and judgment, and we can cordially recommend it to our readers as a useful little pocket companion."—*Engineering.*

A Practical Treatise on Natural and Artificial
Concrete, its Varieties and Constructive Adaptations. By HENRY REID, Author of the 'Science and Art of the Manufacture of Portland Cement.' New Edition, *with 59 woodcuts and 5 plates*, 8vo, cloth, 15s.

Notes on Concrete and Works in Concrete; especially written to assist those engaged upon Public Works. By JOHN NEWMAN, Assoc. Mem. Inst. C.E., crown 8vo, cloth, 4s. 6d.

Electricity as a Motive Power. By Count TH. DU
MONCEL, Membre de l'Institut de France, and FRANK GERALDY, Ingénieur des Ponts et Chaussées. Translated and Edited, with Additions, by C. J. WHARTON, Assoc. Soc. Tel. Eng. and Elec. *With 113 engravings and diagrams*, crown 8vo, cloth, 7s. 6d.

Treatise on Valve-Gears, with special consideration of the Link-Motions of Locomotive Engines. By Dr. GUSTAV ZEUNER, Professor of Applied Mechanics at the Confederated Polytechnikum of Zurich. Translated from the Fourth German Edition, by Professor J. F. KLEIN, Lehigh University, Bethlehem, Pa. *Illustrated*, 8vo, cloth, 12s. 6d.

The French-Polisher's Manual. By a French-Polisher; containing Timber Staining, Washing, Matching, Improving, Painting, Imitations, Directions for Staining, Sizing, Embodying, Smoothing, Spirit Varnishing, French-Polishing, Directions for Re-polishing. Third edition, royal 32mo, sewed, 6d.

Hops, their Cultivation, Commerce, and Uses in
various Countries. By P. L. SIMMONDS. Crown 8vo, cloth, 4s. 6d.

The Principles of Graphic Statics. By GEORGE
SYDENHAM CLARKE, Major Royal Engineers. *With 112 illustrations.* Second edition, 4to, cloth, 12s. 6d.

Dynamo Tenders' Hand-Book. By F. B. BADT, late
1st Lieut. Royal Prussian Artillery. *With* 70 *illustrations.* Third edition,
18mo, cloth, 4s. 6d.

*Practical Geometry, Perspective, and Engineering
Drawing;* a Course of Descriptive Geometry adapted to the Requirements of the Engineering Draughtsman, including the determination of cast shadows and Isometric Projection, each chapter being followed by numerous examples; to which are added rules for Shading, Shade-lining, etc., together with practical instructions as to the Lining, Colouring, Printing, and general treatment of Engineering Drawings, with a chapter on drawing Instruments. By GEORGE S. CLARKE, Capt. R.E. Second edition, *with* 21 *plates.* 2 vols., cloth, 10s. 6d.

The Elements of Graphic Statics. By Professor
KARL VON OTT, translated from the German by G. S. CLARKE, Capt. R.E., Instructor in Mechanical Drawing, Royal Indian Engineering College. *With* 93 *illustrations,* crown 8vo, cloth, 5s.

A Practical Treatise on the Manufacture and Distribution of Coal Gas. By WILLIAM RICHARDS. Demy 4to, with *numerous wood engravings and* 29 *plates,* cloth, 28s.

SYNOPSIS OF CONTENTS :

Introduction — History of Gas Lighting — Chemistry of Gas Manufacture, by Lewis Thompson, Esq., M.R.C.S.—Coal, with Analyses, by J. Paterson, Lewis Thompson, and G. R. Hislop, Esqrs.—Retorts, Iron and Clay—Retort Setting—Hydraulic Main—Condensers—Exhausters—Washers and Scrubbers—Purifiers—Purification—History of Gas Holder—Tanks, Brick and Stone, Composite, Concrete, Cast-iron, Compound Annular Wrought-iron—Specifications—Gas Holders—Station Meter—Governor—Distribution—Mains—Gas Mathematics, or Formulæ for the Distribution of Gas, by Lewis Thompson, Esq.—Services—Consumers' Meters—Regulators—Burners—Fittings—Photometer—Carburization of Gas—Air Gas and Water Gas—Composition of Coal Gas, by Lewis Thompson, Esq.—Analyses of Gas—Influence of Atmospheric Pressure and Temperature on Gas—Residual Products—Appendix—Description of Retort Settings, Buildings, etc., etc.

*The New Formula for Mean Velocity of Discharge
of Rivers and Canals.* By W. R. KUTTER. Translated from articles in the 'Cultur-Ingénieur,' by LOWIS D'A. JACKSON, Assoc. Inst. C.E. 8vo, cloth, 12s. 6d.

*The Practical Millwright and Engineer's Ready
Reckoner;* or Tables for finding the diameter and power of cog-wheels, diameter, weight, and power of shafts, diameter and strength of bolts, etc. By THOMAS DIXON. Fourth edition, 12mo, cloth, 3s.

Tin : Describing the Chief Methods of Mining,
Dressing and Smelting it abroad ; with Notes upon Arsenic, Bismuth and Wolfram. By ARTHUR G. CHARLETON, Mem. American Inst. of Mining Engineers. *With plates,* 8vo, cloth, 12s. 6d.

B 3

Perspective, Explained and Illustrated. By G. S. CLARKE, Capt. R.E. *With illustrations,* 8vo, cloth, 3s. 6d.

Practical Hydraulics; a Series of Rules and Tables for the use of Engineers, etc., etc. By THOMAS BOX. Ninth edition, *numerous plates,* post 8vo, cloth, 5s.

The Essential Elements of Practical Mechanics; based on the Principle of Work, designed for Engineering Students. By OLIVER BYRNE, formerly Professor of Mathematics, College for Civil Engineers. Third edition, *with* 148 *wood engravings,* post 8vo, cloth, 7s. 6d.

CONTENTS:

Chap. 1. How Work is Measured by a Unit, both with and without reference to a Unit of Time—Chap. 2. The Work of Living Agents, the Influence of Friction, and introduces one of the most beautiful Laws of Motion—Chap. 3. The principles expounded in the first and second chapters are applied to the Motion of Bodies—Chap. 4. The Transmission of Work by simple Machines—Chap. 5. Useful Propositions and Rules.

Breweries and Maltings: their Arrangement, Construction, Machinery, and Plant. By G. SCAMELL, F.R.I.B.A. Second edition, revised, enlarged, and partly rewritten. By F. COLYER, M.I.C.E., M.I.M.E. *With* 20 *plates,* 8vo, cloth, 12s. 6d.

A Practical Treatise on the Construction of Horizontal and Vertical Waterwheels, specially designed for the use of operative mechanics. By WILLIAM CULLEN, Millwright and Engineer. *With* 11 *plates.* Second edition, revised and enlarged, small 4to, cloth, 12s. 6d.

A Practical Treatise on Mill-gearing, Wheels, Shafts, Riggers, etc.; for the use of Engineers. By THOMAS BOX. Third edition, *with* 11 *plates.* Crown 8vo, cloth, 7s. 6d.

Mining Machinery: a Descriptive Treatise on the Machinery, Tools, and other Appliances used in Mining. By G. G. ANDRÉ, F.G.S., Assoc. Inst. C.E., Mem. of the Society of Engineers. Royal 4to, uniform with the Author's Treatise on Coal Mining, containing 182 *plates,* accurately drawn to scale, with descriptive text, in 2 vols., cloth, 3l. 12s.

CONTENTS:

Machinery for Prospecting, Excavating, Hauling, and Hoisting—Ventilation—Pumping—Treatment of Mineral Products, including Gold and Silver, Copper, Tin, and Lead, Iron, Coal, Sulphur, China Clay, Brick Earth, etc.

Tables for Setting out Curves for Railways, Canals, Roads, etc., varying from a radius of five chains to three miles. By A. KENNEDY and R. W. HACKWOOD. *Illustrated* 32mo, cloth, 2s. 6d.

Practical Electrical Notes and Definitions for the
use of Engineering Students and Practical Men. By W. PERREN
MAYCOCK, Assoc. M. Inst. E.E., Instructor in Electrical Engineering at
the Pitlake Institute, Croydon, together with the Rules and Regulations
to be observed in Electrical Installation Work. Second edition. Royal
32mo, roan, gilt edges, 4s. 6d., or cloth, red edges, 3s.

The Draughtsman's Handbook of Plan and Map
Drawing; including instructions for the preparation of Engineering,
Architectural, and Mechanical Drawings. *With numerous illustrations
in the text, and* 33 *plates* (15 *printed in colours*). By G. G. ANDRÉ,
F.G.S., Assoc. Inst. C.E. 4to, cloth, 9s.

CONTENTS:

The Drawing Office and its Furnishings—Geometrical Problems—Lines, Dots, and their
Combinations—Colours, Shading, Lettering, Bordering, and North Points—Scales—Plotting
—Civil Engineers' and Surveyors' Plans—Map Drawing—Mechanical and Architectural
Drawing—Copying and Reducing Trigonometrical Formulæ, etc., etc.

The Boiler-maker's and Iron Ship-builder's Companion,
comprising a series of original and carefully calculated tables, of the
utmost utility to persons interested in the iron trades. By JAMES FODEN,
author of 'Mechanical Tables,' etc. Second edition revised, *with illustrations*, crown 8vo, cloth, 5s.

Rock Blasting: a Practical Treatise on the means
employed in Blasting Rocks for Industrial Purposes. By G. G. ANDRÉ,
F.G.S., Assoc. Inst. C.E. *With* 56 *illustrations and* 12 *plates*, 8vo, cloth,
10s. 6d.

Experimental Science: Elementary, Practical, and
Experimental Physics. By GEO. M. HOPKINS. *Illustrated by* 672
engravings. In one large vol., 8vo, cloth, 15s.

A Treatise on Ropemaking as practised in public and
private Rope-yards, with a Description of the Manufacture, Rules, Tables
of Weights, etc., adapted to the Trade, Shipping, Mining, Railways,
Builders, etc. By R. CHAPMAN, formerly foreman to Messrs. Huddart
and Co., Limehouse, and late Master Ropemaker to H.M. Dockyard,
Deptford. Second edition, 12mo, cloth, 3s.

Laxton's Builders' and Contractors' Tables; for the
use of Engineers, Architects, Surveyors, Builders, Land Agents, and
others. Bricklayer, containing 22 tables, with nearly 30,000 calculations.
4to, cloth, 5s.

Laxton's Builders' and Contractors' Tables. Excavator, Earth, Land, Water, and Gas, containing 53 tables, with nearly
24,000 calculations. 4to, cloth, 5s.

Egyptian Irrigation. By W. WILLCOCKS, M.I.C.E., Indian Public Works Department, Inspector of Irrigation, Egypt. With Introduction by Lieut.-Col. J. C. ROSS, R.E., Inspector-General of Irrigation. *With numerous lithographs and wood engravings*, royal 8vo, cloth, 1*l*. 16*s*.

Screw Cutting Tables for Engineers and Machinists, giving the values of the different trains of Wheels required to produce Screws of any pitch, calculated by Lord Lindsay, M.P., F.R.S., F.R.A.S., etc. Cloth, oblong, 2*s*.

Screw Cutting Tables, for the use of Mechanical Engineers, showing the proper arrangement of Wheels for cutting the Threads of Screws of any required pitch, with a Table for making the Universal Gas-pipe Threads and Taps. By W. A. MARTIN, Engineer. Second edition, oblong, cloth, 1*s*., or sewed, 6*d*.

A Treatise on a Practical Method of Designing Slide-Valve Gears by Simple Geometrical Construction, based upon the principles enunciated in Euclid's Elements, and comprising the various forms of Plain Slide-Valve and Expansion Gearing; together with Stephenson's, Gooch's, and Allan's Link-Motions, as applied either to reversing or to variable expansion combinations. By EDWARD J. COWLING WELCH, Memb. Inst. Mechanical Engineers. Crown 8vo, cloth, 6*s*.

Cleaning and Scouring: a Manual for Dyers, Laundresses, and for Domestic Use. By S. CHRISTOPHER. 18mo, sewed, 6*d*.

A Glossary of Terms used in Coal Mining. By WILLIAM STUKELEY GRESLEY, Assoc. Mem. Inst. C.E., F.G.S., Member of the North of England Institute of Mining Engineers. *Illustrated with numerous woodcuts and diagrams,* crown 8vo, cloth, 5*s*.

A Pocket-Book for Boiler Makers and Steam Users, comprising a variety of useful information for Employer and Workman, Government Inspectors, Board of Trade Surveyors, Engineers in charge of Works and Slips, Foremen of Manufactories, and the general Steam-using Public. By MAURICE JOHN SEXTON. Second edition, royal 32mo, roan, gilt edges, 5*s*.

Electrolysis: a Practical Treatise on Nickeling, Coppering, Gilding, Silvering, the Refining of Metals, and the treatment of Ores by means of Electricity. By HIPPOLYTE FONTAINE, translated from the French by J. A. BERLY, C.E., Assoc. S.T.E. *With engravings.* 8vo, cloth, 9*s*.

Barlow's Tables of Squares, Cubes, Square Roots,
Cube Roots, Reciprocals of all Integer Numbers up to 10,000. Post 8vo, cloth, 6s.

A Practical Treatise on the Steam Engine, containing Plans and Arrangements of Details for Fixed Steam Engines, with Essays on the Principles involved in Design and Construction. By ARTHUR RIGG, Engineer, Member of the Society of Engineers and of the Royal Institution of Great Britain. Demy 4to, *copiously illustrated with woodcuts and 96 plates,* in one Volume, half-bound morocco, 2l. 2s.; or cheaper edition, cloth, 25s.

This work is not, in any sense, an elementary treatise, or history of the steam engine, but is intended to describe examples of Fixed Steam Engines without entering into the wide domain of locomotive or marine practice. To this end illustrations will be given of the most recent arrangements of Horizontal, Vertical, Beam, Pumping, Winding, Portable, Semi-portable, Corliss, Allen, Compound, and other similar Engines, by the most eminent Firms in Great Britain and America. The laws relating to the action and precautions to be observed in the construction of the various details, such as Cylinders, Pistons, Piston-rods, Connecting-rods, Cross-heads, Motion-blocks, Eccentrics, Simple, Expansion, Balanced, and Equilibrium Slide-valves, and Valve-gearing will be minutely dealt with. In this connection will be found articles upon the Velocity of Reciprocating Parts and the Mode of Applying the Indicator, Heat and Expansion of Steam Governors, and the like. It is the writer's desire to draw illustrations from every possible source, and give only those rules that present practice deems correct.

A Practical Treatise on the Science of Land and Engineering Surveying, Levelling, Estimating Quantities, etc., with a general description of the several Instruments required for Surveying, Levelling, Plotting, etc. By H. S. MERRETT. Fourth edition, revised by G. W. USILL, Assoc. Mem. Inst. C.E. 41 *plates, with illustrations and tables,* royal 8vo, cloth, 12s. 6d.

PRINCIPAL CONTENTS:

Part 1. Introduction and the Principles of Geometry. Part 2. Land Surveying; comprising General Observations—The Chain—Offsets Surveying by the Chain only—Surveying Hilly Ground—To Survey an Estate or Parish by the Chain only—Surveying with the Theodolite—Mining and Town Surveying—Railroad Surveying—Mapping—Division and Laying out of Land—Observations on Enclosures—Plane Trigonometry. Part 3. Levelling—Simple and Compound Levelling—The Level Book—Parliamentary Plan and Section—Levelling with a Theodolite—Gradients—Wooden Curves—To Lay out a Railway Curve—Setting out Widths. Part 4. Calculating Quantities generally for Estimates—Cuttings and Embankments—Tunnels—Brickwork—Ironwork—Timber Measuring. Part 5. Description and Use of Instruments in Surveying and Plotting—The Improved Dumpy Level—Troughton's Level—The Prismatic Compass—Proportional Compass—Box Sextant—Vernier—Pantagraph—Merrett's Improved Quadrant—Improved Computation Scale—The Diagonal Scale—Straight Edge and Sector. Part 6. Logarithms of Numbers—Logarithmic Sines and Co-Sines, Tangents and Co-Tangents—Natural Sines and Co-Sines—Tables for Earthwork, for Setting out Curves, and for various Calculations, etc., etc., etc.

Mechanical Graphics. A Second Course of Mechanical Drawing. With Preface by Prof. PERRY, B.Sc., F.R.S. Arranged for use in Technical and Science and Art Institutes, Schools and Colleges, by GEORGE HALLIDAY, Whitworth Scholar. 8vo, cloth, 6s.

The Assayer's Manual: an Abridged Treatise on the Docimastic Examination of Ores and Furnace and other Artificial Products. By BRUNO KERL. Translated by W. T. BRANNT. *With 65 illustrations,* 8vo, cloth, 12s. 6d.

Dynamo-Electric Machinery: a Text-Book for Students of Electro-Technology. By SILVANUS P. THOMPSON, B.A., D.Sc., M.S.T.E. [*New edition in the press.*

The Practice of Hand Turning in Wood, Ivory, Shell, etc., with Instructions for Turning such Work in Metal as may be required in the Practice of Turning in Wood, Ivory, etc.; also an Appendix on Ornamental Turning. (A book for beginners.) By FRANCIS CAMPIN. Third edition, *with wood engravings,* crown 8vo, cloth, 6s.

CONTENTS:

On Lathes—Turning Tools—Turning Wood—Drilling—Screw Cutting—Miscellaneous Apparatus and Processes—Turning Particular Forms—Staining—Polishing—Spinning Metals —Materials—Ornamental Turning, etc.

Treatise on Watchwork, Past and Present. By the Rev. H. L. NELTHROPP, M.A., F.S.A. *With 32 illustrations,* crown 8vo, cloth, 6s. 6d.

CONTENTS:

Definitions of Words and Terms used in Watchwork—Tools—Time—Historical Summary—On Calculations of the Numbers for Wheels and Pinions; their Proportional Sizes, Trains, etc.—Of Dial Wheels, or Motion Work—Length of Time of Going without Winding up—The Verge—The Horizontal—The Duplex—The Lever—The Chronometer—Repeating Watches—Keyless Watches—The Pendulum, or Spiral Spring—Compensation—Jewelling of Pivot Holes—Clerkenwell—Fallacies of the Trade—Incapacity of Workmen—How to Choose and Use a Watch, etc.

Algebra Self-Taught. By W. P. HIGGS, M.A., D.Sc., LL.D., Assoc. Inst. C.E., Author of 'A Handbook of the Differential Calculus,' etc. Second edition, crown 8vo, cloth, 2s. 6d.

CONTENTS:

Symbols and the Signs of Operation—The Equation and the Unknown Quantity—Positive and Negative Quantities—Multiplication—Involution—Exponents—Negative Exponents—Roots, and the Use of Exponents as Logarithms—Tables of Logarithms and Proportionate Parts—Transformation of System of Logarithms—Common Uses of Common Logarithms—Compound Multiplication and the Binomial Theorem—Division, Fractions, and Ratio—Continued Proportion—The Series and the Summation of the Series—Limit of Series—Square and Cube Roots—Equations—List of Formulæ, etc.

Spons' Dictionary of Engineering, Civil, Mechanical, Military, and Naval; with technical terms in French, German, Italian, and Spanish, 3100 pp., and *nearly* 8000 *engravings,* in super-royal 8vo, in 8 divisions, 5l. 8s. Complete in 3 vols., cloth, 5l. 5s. Bound in a superior manner, half-morocco, top edge gilt, 3 vols., 6l. 12s.

Notes in Mechanical Engineering. Compiled principally for the use of the Students attending the Classes on this subject at the City of London College. By HENRY ADAMS, Mem. Inst. M.E., Mem. Inst. C.E., Mem. Soc. of Engineers. Crown 8vo, cloth, 2s. 6d.

Canoe and Boat Building: a complete Manual for Amateurs, containing plain and comprehensive directions for the construction of Canoes, Rowing and Sailing Boats, and Hunting Craft. By W. P. STEPHENS. *With numerous illustrations and* 24 *plates of Working Drawings.* Crown 8vo, cloth, 9s.

Proceedings of the National Conference of Electricians, Philadelphia, October 8th to 13th, 1884. 18mo, cloth, 3s.

Dynamo - Electricity, its Generation, Application, Transmission, Storage, and Measurement. By G. B. PRESCOTT. *With* 545 *illustrations.* 8vo, cloth, 1l. 1s.

Domestic Electricity for Amateurs. Translated from the French of E. HOSPITALIER, Editor of "L'Electricien," by C. J. WHARTON, Assoc. Soc. Tel. Eng. *Numerous illustrations.* Demy 8vo, cloth, 6s.

CONTENTS:

1. Production of the Electric Current—2. Electric Bells—3. Automatic Alarms—4. Domestic Telephones—5. Electric Clocks—6. Electric Lighters—7. Domestic Electric Lighting—8. Domestic Application of the Electric Light—9. Electric Motors—10. Electrical Locomotion—11. Electrotyping, Plating, and Gilding—12. Electric Recreations—13. Various applications—Workshop of the Electrician.

Wrinkles in Electric Lighting. By VINCENT STEPHEN. *With illustrations.* 18mo, cloth, 2s. 6d.

CONTENTS:

1. The Electric Current and its production by Chemical means—2. Production of Electric Currents by Mechanical means—3. Dynamo-Electric Machines—4. Electric Lamps—5. Lead—6. Ship Lighting.

Foundations and Foundation Walls for all classes of Buildings, Pile Driving, Building Stones and Bricks, Pier and Wall construction, Mortars, Limes, Cements, Concretes, Stuccos, &c. 64 *illustrations.* By G. T. POWELL and F. BAUMAN. 8vo, cloth, 10s. 6d.

Manual for Gas Engineering Students. By D. LEE. 18mo, cloth, 1s.

Telephones, their Construction and Management.
By F. C. ALLSOP. Crown 8vo, cloth, 5s.

Hydraulic Machinery, Past and Present. A Lecture delivered to the London and Suburban Railway Officials' Association. By H. ADAMS, Mem. Inst. C.E. *Folding plate.* 8vo, sewed, 1s.

Twenty Years with the Indicator. By THOMAS PRAY, Jun., C.E., M.E., Member of the American Society of Civil Engineers. 2 vols., royal 8vo, cloth, 12s. 6d.

Annual Statistical Report of the Secretary to the Members of the Iron and Steel Association on the Home and Foreign Iron and Steel Industries in 1889. Issued June 1890. 8vo, sewed, 5s.

Bad Drains, and How to Test them; with Notes on the Ventilation of Sewers, Drains, and Sanitary Fittings, and the Origin and Transmission of Zymotic Disease. By R. HARRIS REEVES. Crown 8vo, cloth, 3s. 6d.

Well Sinking. The modern practice of Sinking and Boring Wells, with geological considerations and examples of Wells. By ERNEST SPON, Assoc. Mem. Inst. C.E., Mem. Soc. Eng., and of the Franklin Inst., etc. Second edition, revised and enlarged. Crown 8vo, cloth, 10s. 6d.

The Voltaic Accumulator: an Elementary Treatise. By ÉMILE REYNIER. Translated by J. A. BERLY, Assoc. Inst. E.E. *With* 62 *illustrations,* 8vo, cloth, 9s.

Ten Years' Experience in Works of Intermittent Downward Filtration. By J. BAILEY DENTON, Mem. Inst. C.E. Second edition, with additions. Royal 8vo, cloth, 5s.

Land Surveying on the Meridian and Perpendicular System. By WILLIAM PENMAN, C.E. 8vo, cloth, 8s. 6d.

The Electromagnet and Electromagnetic Mechanism. By SILVANUS P. THOMPSON, D.Sc., F.R.S. Second edition, 8vo, cloth, 15s.

Incandescent Wiring Hand-Book. By F. B. BADT,
late 1st Lieut. Royal Prussian Artillery. *With* 41 *illustrations and* 5 *tables.* 18mo, cloth, 4s. 6d.

A Pocket-book for Pharmacists, Medical Prac-
titioners, Students, etc., etc. (*British, Colonial, and American*). By THOMAS BAYLEY, Assoc. R. Coll. of Science, Consulting Chemist, Analyst, and Assayer, Author of a 'Pocket-book for Chemists,' 'The Assay and Analysis of Iron and Steel, Iron Ores, and Fuel,' etc., etc. Royal 32mo, boards, gilt edges, 6s.

The Fireman's Guide; a Handbook on the Care of
Boilers. By TEKNOLOG, föreningen T. I. Stockholm. Translated from the third edition, and revised by KARL P. DAHLSTROM, M.E. Second edition. Fcap. 8vo, cloth, 2s.

The Mechanician: A Treatise on the Construction
and Manipulation of Tools, for the use and instruction of Young Engineers and Scientific Amateurs, comprising the Arts of Blacksmithing and Forging; the Construction and Manufacture of Hand Tools, and the various Methods of Using and Grinding them; description of Hand and Machine Processes; Turning and Screw Cutting. By CAMERON KNIGHT, Engineer. *Containing* 1147 *illustrations,* and 397 pages of letter-press. Fourth edition, 4to, cloth, 18s.

A Treatise on Modern Steam Engines and Boilers,
including Land Locomotive, and Marine Engines and Boilers, for the use of Students. By FREDERICK COLYER, M. Inst. C.E., Mem. Inst. M.E. With 36 *plates.* 4to, cloth, 12s. 6d.

CONTENTS:

1. Introduction—2. Original Engines—3. Boilers—4. High-Pressure Beam Engines—5. Cornish Beam Engines—6. Horizontal Engines—7. Oscillating Engines—8. Vertical High-Pressure Engines—9. Special Engines—10. Portable Engines—11. Locomotive Engines—12. Marine Engines.

Steam Engine Management; a Treatise on the
Working and Management of Steam Boilers. By F. COLYER, M. Inst. C.E., Mem. Inst. M.E. New edition, 18mo, cloth, 3s. 6d.

Aid Book to Engineering Enterprise. By EWING
MATHESON, M. Inst. C.E. The Inception of Public Works, Parliamentary Procedure for Railways, Concessions for Foreign Works, and means of Providing Money, the Points which determine Success or Failure, Contract and Purchase, Commerce in Coal, Iron, and Steel, &c. Second edition, revised and enlarged, 8vo, cloth, 21s.

Pumps, Historically, Theoretically, and Practically
Considered. By P. R. BJÖRLING. With 156 *illustrations*. Crown 8vo, cloth, 7s. 6d.

The Marine Transport of Petroleum. A Book for the use of Shipowners, Shipbuilders, Underwriters, Merchants, Captains and Officers of Petroleum-carrying Vessels. By G. H. LITTLE, Editor of the 'Liverpool Journal of Commerce.' Crown 8vo, cloth, 10s. 6d.

Liquid Fuel for Mechanical and Industrial Purposes.
Compiled by E. A. BRAYLEY HODGETTS. *With wood engravings.* 8vo, cloth, 7s. 6d.

Tropical Agriculture: A Treatise on the Culture, Preparation, Commerce and Consumption of the principal Products of the Vegetable Kingdom. By P. L. SIMMONDS, F.L.S., F.R.C.I. New edition, revised and enlarged, 8vo, cloth, 21s.

Health and Comfort in House Building; or, Ventilation with Warm Air by Self-acting Suction Power. With Review of the Mode of Calculating the Draught in Hot-air Flues, and with some Actual Experiments by J. DRYSDALE, M.D., and J. W. HAYWARD, M.D. *With plates and woodcuts.* Third edition, with some New Sections, and the whole carefully Revised, 8vo, cloth, 7s. 6d.

Losses in Gold Amalgamation. With Notes on the Concentration of Gold and Silver Ores. *With six plates.* By W. McDERMOTT and P. W. DUFFIELD. 8vo, cloth, 5s.

A Guide for the Electric Testing of Telegraph Cables.
By Col. V. HOSKIŒR, Royal Danish Engineers. Third edition, crown 8vo, cloth, 4s. 6d.

The Hydraulic Gold Miners' Manual. By T. S. G. KIRKPATRICK, M.A. Oxon. *With 6 plates.* Crown 8vo, cloth, 6s.

"We venture to think that this work will become a text-book on the important subject of which it treats. Until comparatively recently hydraulic mines were neglected. This was scarcely to be surprised at, seeing that their working in California was brought to an abrupt termination by the action of the farmers on the *débris* question, whilst their working in other parts of the world had not been attended with the anticipated success."—*The Mining World and Engineering Record.*

A Text-Book of Tanning, embracing the Preparation of all kinds of Leather. By HARRY R. PROCTOR, F.C.S., of Low Lights Tanneries. *With illustrations.* Crown 8vo, cloth, 10s. 6d.

The Arithmetic of Electricity. By T. O'CONOR SLOANE. Crown 8vo, cloth, 4s. 6d.

The Turkish Bath: Its Design and Construction for Public and Commercial Purposes. By R. O. ALLSOP, Architect. *With plans and sections.* 8vo, cloth, 6s.

Earthwork Slips and Subsidences upon Public Works: Their Causes, Prevention and Reparation. Especially written to assist those engaged in the Construction or Maintenance of Railways, Docks, Canals, Waterworks, River Banks, Reclamation Embankments, Drainage Works, &c., &c. By JOHN NEWMAN, Assoc. Mem. Inst. C.E., Author of 'Notes on Concrete,' &c. Crown 8vo, cloth, 7s. 6d.

Gas and Petroleum Engines: A Practical Treatise on the Internal Combustion Engine. By WM. ROBINSON, M.E., Senior Demonstrator and Lecturer on Applied Mechanics, Physics, &c., City and Guilds of London College, Finsbury, Assoc. Mem. Inst. C.E., &c. *Numerous illustrations.* 8vo, cloth, 14s.

Waterways and Water Transport in Different Countries. With a description of the Panama, Suez, Manchester, Nicaraguan, and other Canals. By J. STEPHEN JEANS, Author of 'England's Supremacy,' 'Railway Problems,' &c. *Numerous illustrations.* 8vo, cloth, 14s.

A Treatise on the Richards Steam-Engine Indicator and the Development and Application of Force in the Steam-Engine. By CHARLES T. PORTER. Fourth Edition, revised and enlarged, 8vo, cloth, 9s.

CONTENTS.

The Nature and Use of the Indicator: The several lines on the Diagram. Examination of Diagram No. 1.
Of Truth in the Diagram.
Description of the Richards Indicator.
Practical Directions for Applying and Taking Care of the Indicator.
Introductory Remarks.
Units.
Expansion.
Directions for ascertaining from the Diagram the Power exerted by the Engine.
To Measure from the Diagram the Quantity of Steam Consumed.
To Measure from the Diagram the Quantity of Heat Expended.
Of the Real Diagram, and how to Construct it.
Of the Conversion of Heat into Work in the Steam-engine.
Observations on the several Lines of the Diagram.
Of the Loss attending the Employment of Slow-piston Speed, and the Extent to which this is Shown by the Indicator.
Of other Applications of the Indicator.
Of the use of the Tables of the Properties of Steam in Calculating the Duty of Boilers.
Introductory.
Of the Pressure on the Crank when the Connecting-rod is conceived to be of Infinite Length.
The Modification of the Acceleration and Retardation that is occasioned by the Angular Vibration of the Connecting-rod.
Method of representing the actual pressure on the crank at every point of its revolution.
The Rotative Effect of the Pressure exerted on the Crank.
The Transmitting Parts of an Engine, considered as an Equaliser of Motion.
A Ride on a Buffer-beam (Appendix).

In demy 4to, handsomely bound in cloth, *illustrated with* **220** *full page plates*,
Price 15*s*.

ARCHITECTURAL EXAMPLES
IN BRICK, STONE, WOOD, AND IRON.
A COMPLETE WORK ON THE DETAILS AND ARRANGEMENT OF BUILDING CONSTRUCTION AND DESIGN.

By WILLIAM FULLERTON, Architect.

Containing 220 Plates, with numerous Drawings selected from the Architecture of Former and Present Times.

The Details and Designs are Drawn to Scale, $\frac{1}{8}"$, $\frac{1}{4}"$, $\frac{1}{2}"$, and Full size being chiefly used.

The Plates are arranged in Two Parts. The First Part contains Details of Work in the four principal Building materials, the following being a few of the subjects in this Part:—Various forms of Doors and Windows, Wood and Iron Roofs, Half Timber Work, Porches, Towers, Spires, Belfries, Flying Buttresses, Groining, Carving, Church Fittings, Constructive and Ornamental Iron Work, Classic and Gothic Molds and Ornament, Foliation Natural and Conventional, Stained Glass, Coloured Decoration, a Section to Scale of the Great Pyramid, Grecian and Roman Work, Continental and English Gothic, Pile Foundations, Chimney Shafts according to the regulations of the London County Council, Board Schools. The Second Part consists of Drawings of Plans and Elevations of Buildings, arranged under the following heads:—Workmen's Cottages and Dwellings, Cottage Residences and Dwelling Houses, Shops, Factories, Warehouses, Schools, Churches and Chapels, Public Buildings, Hotels and Taverns, and Buildings of a general character.

All the Plates are accompanied with particulars of the Work, with Explanatory Notes and Dimensions of the various parts.

Specimen Pages, reduced from the originals.

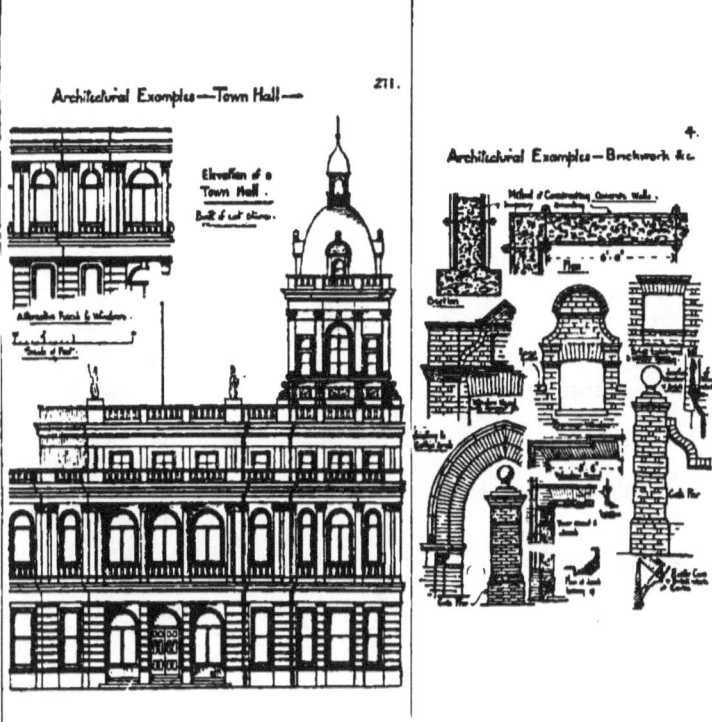

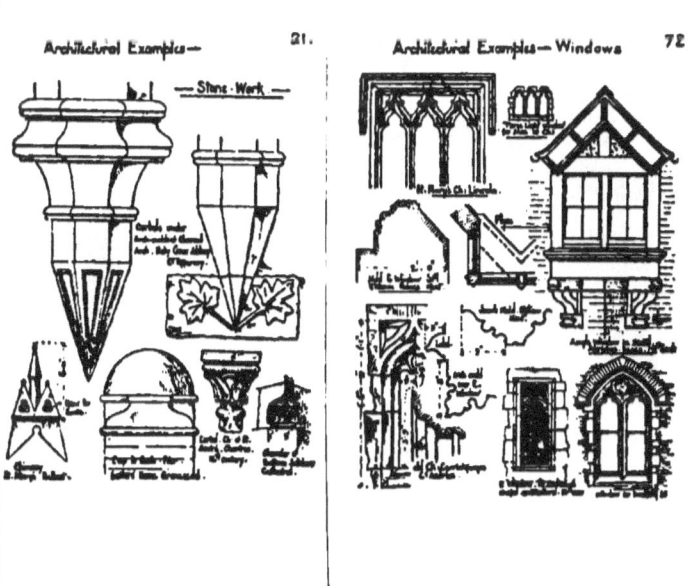

Crown 8vo, cloth, with illustrations, 5s.

WORKSHOP RECEIPTS,

FIRST SERIES.

By ERNEST SPON.

SYNOPSIS OF CONTENTS.

Bookbinding.
Bronzes and Bronzing.
Candles.
Cement.
Cleaning.
Colourwashing.
Concretes.
Dipping Acids.
Drawing Office Details.
Drying Oils.
Dynamite.
Electro - Metallurgy — (Cleaning, Dipping, Scratch-brushing, Batteries, Baths, and Deposits of every description).
Enamels.
Engraving on Wood, Copper, Gold, Silver, Steel, and Stone.
Etching and Aqua Tint.
Firework Making — (Rockets, Stars, Rains, Gerbes, Jets, Tourbillons, Candles, Fires, Lances, Lights, Wheels, Fire-balloons, and minor Fireworks).
Fluxes.
Foundry Mixtures.

Freezing.
Fulminates.
Furniture Creams, Oils, Polishes, Lacquers, and Pastes.
Gilding.
Glass Cutting, Cleaning, Frosting, Drilling, Darkening, Bending, Staining, and Painting.
Glass Making.
Glues.
Gold.
Graining.
Gums.
Gun Cotton.
Gunpowder.
Horn Working.
Indiarubber.
Japans, Japanning, and kindred processes.
Lacquers.
Lathing.
Lubricants.
Marble Working.
Matches.
Mortars.
Nitro-Glycerine.
Oils.

Paper.
Paper Hanging.
Painting in Oils, in Water Colours, as well as Fresco, House, Transparency, Sign, and Carriage Painting.
Photography.
Plastering.
Polishes.
Pottery—(Clays, Bodies, Glazes, Colours, Oils, Stains, Fluxes, Enamels, and Lustres).
Scouring.
Silvering.
Soap.
Solders.
Tanning.
Taxidermy.
Tempering Metals.
Treating Horn, Mother-o'-Pearl, and like substances.
Varnishes, Manufacture and Use of.
Veneering.
Washing.
Waterproofing.
Welding.

Besides Receipts relating to the lesser Technological matters and processes, such as the manufacture and use of Stencil Plates, Blacking, Crayons, Paste, Putty, Wax, Size, Alloys, Catgut, Tunbridge Ware, Picture Frame and Architectural Mouldings, Compos, Cameos, and others too numerous to mention.

Crown 8vo, cloth, 485 pages, with illustrations, 5s.

WORKSHOP RECEIPTS,

SECOND SERIES.

By ROBERT HALDANE.

SYNOPSIS OF CONTENTS.

Acidimetry and Alkalimetry.	Disinfectants.	Iodoform.
Albumen.	Dyeing, Staining, and Colouring.	Isinglass.
Alcohol.	Essences.	Ivory substitutes.
Alkaloids.	Extracts.	Leather.
Baking-powders.	Fireproofing.	Luminous bodies.
Bitters.	Gelatine, Glue, and Size.	Magnesia.
Bleaching.	Glycerine.	Matches.
Boiler Incrustations.	Gut.	Paper.
Cements and Lutes.	Hydrogen peroxide.	Parchment.
Cleansing.	Ink.	Perchloric acid.
Confectionery.	Iodine.	Potassium oxalate.
Copying.		Preserving.

Pigments, Paint, and Painting: embracing the preparation of *Pigments*, including alumina lakes, blacks (animal, bone, Frankfort, ivory, lamp, sight, soot), blues (antimony, Antwerp, cobalt, cæruleum, Egyptian, manganate, Paris, Péligot, Prussian, smalt, ultramarine), browns (bistre, hinau, sepia, sienna, umber, Vandyke), greens (baryta, Brighton, Brunswick, chrome, cobalt, Douglas, emerald, manganese, mitis, mountain, Prussian, sap, Scheele's, Schweinfurth, titanium, verdigris, zinc), reds (Brazilwood lake, carminated lake, carmine, Cassius purple, cobalt pink, cochineal lake, colcothar, Indian red, madder lake, red chalk, red lead, vermilion), whites (alum, baryta, Chinese, lead sulphate, white lead—by American, Dutch, French, German, Kremnitz, and Pattinson processes, precautions in making, and composition of commercial samples—whiting, Wilkinson's white, zinc white), yellows (chrome, gamboge, Naples, orpiment, realgar, yellow lakes); *Paint* (vehicles, testing oils, driers, grinding, storing, applying, priming, drying, filling, coats, brushes, surface, water-colours, removing smell, discoloration; miscellaneous paints—cement paint for carton-pierre, copper paint, gold paint, iron paint, lime paints, silicated paints, steatite paint, transparent paints, tungsten paints, window paint, zinc paints); *Painting* (general instructions, proportions of ingredients, measuring paint work; carriage painting—priming paint, best putty, finishing colour, cause of cracking, mixing the paints, oils, driers, and colours, varnishing, importance of washing vehicles, re-varnishing, how to dry paint; woodwork painting).

Crown 8vo, cloth, 480 pages, with 183 illustrations, 5s.

WORKSHOP RECEIPTS,

THIRD SERIES.

By C. G. WARNFORD LOCK.

Uniform with the First and Second Series.

Synopsis of Contents.

Alloys.	Indium.	Rubidium.
Aluminium.	Iridium.	Ruthenium.
Antimony.	Iron and Steel.	Selenium.
Barium.	Lacquers and Lacquering.	Silver.
Beryllium.	Lanthanum.	Slag.
Bismuth.	Lead.	Sodium.
Cadmium.	Lithium.	Strontium.
Cæsium.	Lubricants.	Tantalum.
Calcium.	Magnesium.	Terbium.
Cerium.	Manganese.	Thallium.
Chromium.	Mercury.	Thorium.
Cobalt.	Mica.	Tin.
Copper.	Molybdenum.	Titanium.
Didymium.	Nickel.	Tungsten.
Electrics.	Niobium.	Uranium.
Enamels and Glazes.	Osmium.	Vanadium.
Erbium.	Palladium.	Yttrium.
Gallium.	Platinum.	Zinc.
Glass.	Potassium.	Zirconium.
Gold.	Rhodium.	

WORKSHOP RECEIPTS,
FOURTH SERIES,
DEVOTED MAINLY TO HANDICRAFTS & MECHANICAL SUBJECTS.

By C. G. WARNFORD LOCK.

250 Illustrations, with Complete Index, and a General Index to the Four Series, 5s.

Waterproofing — rubber goods, cuprammonium processes, miscellaneous preparations.

Packing and Storing articles of delicate odour or colour, of a deliquescent character, liable to ignition, apt to suffer from insects or damp, or easily broken.

Embalming and Preserving anatomical specimens.

Leather Polishes:

Cooling Air and Water, producing low temperatures, making ice, cooling syrups and solutions, and separating salts from liquors by refrigeration.

Pumps and Siphons, embracing every useful contrivance for raising and supplying water on a moderate scale, and moving corrosive, tenacious, and other liquids.

Desiccating—air- and water-ovens, and other appliances for drying natural and artificial products.

Distilling—water, tinctures, extracts, pharmaceutical preparations, essences, perfumes, and alcoholic liquids.

Emulsifying as required by pharmacists and photographers.

Evaporating—saline and other solutions, and liquids demanding special precautions.

Filtering—water, and solutions of various kinds.

Percolating and Macerating.

Electrotyping.

Stereotyping by both plaster and paper processes.

Bookbinding in all its details.

Straw Plaiting and the fabrication of baskets, matting, etc.

Musical Instruments—the preservation, tuning, and repair of pianos, harmoniums, musical boxes, etc.

Clock and Watch Mending—adapted for intelligent amateurs.

Photography—recent development in rapid processes, handy apparatus, numerous recipes for sensitizing and developing solutions, and applications to modern illustrative purposes.

Crown 8vo, cloth, with 373 illustrations, price 5s.

WORKSHOP RECEIPTS,

FIFTH SERIES.

By C. G. WARNFORD LOCK, F.L.S.

Containing many new Articles, as well as additions to Articles included in the previous Series, as follows, viz. :—

Anemometers.
Barometers, How to make.
Boat Building.
Camera Lucida, How to use.
Cements and Lutes.
Cooling.
Copying.
Corrosion and Protection of Metal Surfaces.
Dendrometer, How to use.
Desiccating.
Diamond Cutting and Polishing. Electrics. New Chemical Batteries, Bells, Commutators, Galvanometers, Cost of Electric Lighting, Microphones, Simple Motors, Phonogram and Graphophone, Registering Apparatus, Regulators, Electric Welding and Apparatus, Transformers.
Evaporating.
Explosives.
Filtering.
Fireproofing, Buildings, Textile Fabrics.
Fire-extinguishing Compounds and Apparatus.
Glass Manipulating. Drilling, Cutting, Breaking, Etching, Frosting, Powdering, &c.
Glass Manipulations for Laboratory Apparatus.
Labels. Lacquers.
Illuminating Agents.
Inks. Writing, Copying, Invisible, Marking, Stamping.
Magic Lanterns, their management and preparation of slides.
Metal Work. Casting Ornamental Metal Work, Copper Welding, Enamels for Iron and other Metals, Gold Beating, Smiths' Work.
Modelling and Plaster Casting.
Netting.
Packing and Storing. Acids, &c.
Percolation.
Preserving Books.
Preserving Food, Plants, &c.
Pumps and Syphons for various liquids.
Repairing Books.
Rope Tackle.
Stereotyping.
Taps, Various.
Tobacco Pipe Manufacture.
Tying and Splicing Ropes.
Velocipedes, Repairing.
Walking Sticks.
Waterproofing.

NOW COMPLETE.

With nearly 1500 *illustrations*, in super-royal 8vo, in 5 Divisions, cloth. Divisions 1 to 4, 13*s*. 6*d*. each ; Division 5, 17*s*. 6*d*. ; or 2 vols., cloth, £3 10*s*.

SPONS' ENCYCLOPÆDIA
OF THE
INDUSTRIAL ARTS, MANUFACTURES, AND COMMERCIAL PRODUCTS.

EDITED BY C. G. WARNFORD LOCK, F.L.S.

Among the more important of the subjects treated of, are the following :—

Acids, 207 pp. 220 figs.
Alcohol, 23 pp. 16 figs.
Alcoholic Liquors, 13 pp.
Alkalies, 89 pp. 78 figs.
Alloys.   Alum.
Asphalt.   Assaying.
Beverages, 89 pp. 29 figs.
Blacks.
Bleaching Powder, 15 pp.
Bleaching, 51 pp. 48 figs.
Candles, 18 pp. 9 figs.
Carbon Bisulphide.
Celluloid, 9 pp.
Cements.   Clay.
Coal-tar Products, 44 pp. 14 figs.
Cocoa, 8 pp.
Coffee, 32 pp. 13 figs.
Cork, 8 pp. 17 figs.
Cotton Manufactures, 62 pp. 57 figs.
Drugs, 38 pp.
Dyeing and Calico Printing, 28 pp. 9 figs.
Dyestuffs, 16 pp.
Electro-Metallurgy, 13 pp.
Explosives, 22 pp. 33 figs.
Feathers.
Fibrous Substances, 92 pp. 79 figs.
Floor-cloth, 16 pp. 21 figs.
Food Preservation, 8 pp.
Fruit, 8 pp.

Fur, 5 pp.
Gas, Coal, 8 pp.
Gems.
Glass, 45 pp. 77 figs.
Graphite, 7 pp.
Hair, 7 pp.
Hair Manufactures.
Hats, 26 pp. 26 figs.
Honey.   Hops.
Horn.
Ice, 10 pp. 14 figs.
Indiarubber Manufactures, 23 pp. 17 figs.
Ink, 17 pp.
Ivory.
Jute Manufactures, 11 pp., 11 figs.
Knitted Fabrics — Hosiery, 15 pp. 13 figs.
Lace, 13 pp. 9 figs.
Leather, 28 pp. 31 figs.
Linen Manufactures, 16 pp. 6 figs.
Manures, 21 pp. 30 figs.
Matches, 17 pp. 38 figs.
Mordants, 13 pp.
Narcotics, 47 pp.
Nuts, 10 pp.
Oils and Fatty Substances, 125 pp.
Paint.
Paper, 26 pp. 23 figs.
Paraffin, 8 pp. 6 figs.
Pearl and Coral, 8 pp.
Perfumes, 10 pp.

Photography, 13 pp. 20 figs.
Pigments, 9 pp. 6 figs.
Pottery, 46 pp. 57 figs.
Printing and Engraving, 20 pp. 8 figs.
Rags.
Resinous and Gummy Substances, 75 pp. 16 figs.
Rope, 16 pp. 17 figs.
Salt, 31 pp. 23 figs.
Silk, 8 pp.
Silk Manufactures, 9 pp. 11 figs.
Skins, 5 pp.
Small Wares, 4 pp.
Soap and Glycerine, 39 pp. 45 figs.
Spices, 16 pp.
Sponge, 5 pp.
Starch, 9 pp. 10 figs.
Sugar, 155 pp. 134 figs.
Sulphur.
Tannin, 18 pp.
Tea, 12 pp.
Timber, 13 pp.
Varnish, 15 pp.
Vinegar, 5 pp.
Wax, 5 pp.
Wool, 2 pp.
Woollen Manufactures, 58 pp. 39 figs.

MECHANICAL MANIPULATION.

THE MECHANICIAN:
A TREATISE ON THE CONSTRUCTION AND MANIPULATION OF TOOLS, FOR THE USE AND INSTRUCTION OF YOUNG ENGINEERS AND SCIENTIFIC AMATEURS;

Comprising the Arts of Blacksmithing and Forging; the Construction and Manufacture of Hand Tools, and the various Methods of Using and Grinding them; the Construction of Machine Tools, and how to work them; Turning and Screw-cutting; the various details of setting out work, &c., &c.

By CAMERON KNIGHT, Engineer.

96 4to plates, containing 1147 illustrations, and 397 pages of letterpress, second edition, reprinted from the first, 4to, cloth, **18s.**

Of the six chapters constituting the work, the first is devoted to forging; in which the fundamental principles to be observed in making forged articles of every class are stated, giving the proper relative positions for the constituent fibres of each article, the mode of selecting proper quantities of material, steam-hammer operations, shaping-moulds, and the manipulations resorted to for shaping the component masses to the intended forms.

Engineers' tools and their construction are next treated, because they must be used during all operations described in the remaining chapters, the author thinking that the student should first acquire knowledge of the apparatus which he is supposed to be using in the course of the processes given in Chapters 4, 5, and 6. In the fourth chapter planing and lining are treated, because these are the elements of machine-making in general. The processes described in this chapter are those on which all accuracy of fitting and finishing depend. The next chapter, which treats of shaping and slotting, the author endeavours to render comprehensive by giving the hand-shaping processes in addition to the machine-shaping.

In many cases hand-shaping is indispensable, such as sudden breakage, operations abroad, and on board ship, also for constructors having a limited number of machines. Turning and screw-cutting occupy the last chapter. In this, the operations for lining, centering, turning, and screw-forming are detailed and their principles elucidated.

The Mechanician is the result of the author's experience in engine making during twenty years; and he has concluded that, however retentive the memory of a machinist might be, it would be convenient for him to have a book of primary principles and processes to which he could refer with confidence.

JUST PUBLISHED.

In demy 8vo, cloth, 600 pages, and 1420 Illustrations, 6s.

SPONS'
MECHANICS' OWN BOOK;
A MANUAL FOR HANDICRAFTSMEN AND AMATEURS.

CONTENTS.

Mechanical Drawing—Casting and Founding in Iron, Brass, Bronze, and other Alloys—Forging and Finishing Iron—Sheetmetal Working—Soldering, Brazing, and Burning—Carpentry and Joinery, embracing descriptions of some 400 Woods, over 200 Illustrations of Tools and their uses, Explanations (with Diagrams) of 116 joints and hinges, and Details of Construction of Workshop appliances, rough furniture, Garden and Yard Erections, and House Building—Cabinet-Making and Veneering—Carving and Fretcutting—Upholstery—Painting, Graining, and Marbling—Staining Furniture, Woods, Floors, and Fittings—Gilding, dead and bright, on various grounds—Polishing Marble, Metals, and Wood—Varnishing—Mechanical movements, illustrating contrivances for transmitting motion—Turning in Wood and Metals—Masonry, embracing Stonework, Brickwork, Terracotta and Concrete—Roofing with Thatch, Tiles, Slates, Felt, Zinc, &c.—Glazing with and without putty, and lead glazing—Plastering and Whitewashing—Paper-hanging—Gas-fitting—Bell-hanging, ordinary and electric Systems—Lighting—Warming--Ventilating—Roads, Pavements, and Bridges—Hedges, Ditches, and Drains—Water Supply and Sanitation—Hints on House Construction suited to new countries.

E. & F. N. SPON, 125, Strand, London.
New York: 12, Cortlandt Street.

SPONS' DICTIONARY OF ENGINEERING,

CIVIL, MECHANICAL, MILITARY, & NAVAL,

WITH

Technical Terms in French, German, Italian, and Spanish.

In 97 numbers, Super-royal 8vo, containing 3132 *printed pages* and 7414 *engravings*. Any number can be had separate: Nos. 1 to 95 1s. each, post free; Nos. 96, 97, 2s., post free. *See also page* 112.

COMPLETE LIST OF ALL THE SUBJECTS:

	Nos.		Nos.
Abacus	1	Barrage	8 and 9
Adhesion	1	Battery	9 and 10
Agricultural Engines	1 and 2	Bell and Bell-hanging	10
Air-Chamber	2	Belts and Belting	10 and 11
Air-Pump	2	Bismuth	11
Algebraic Signs	2	Blast Furnace	11 and 12
Alloy	2	Blowing Machine	12
Aluminium	2	Body Plan	12 and 13
Amalgamating Machine	2	Boilers	13, 14, 15
Ambulance	2	Bond	15 and 16
Anchors	2	Bone Mill	16
Anemometer	2 and 3	Boot-making Machinery	16
Angular Motion	3 and 4	Boring and Blasting	16 to 19
Angle-iron	3	Brake	19 and 20
Angle of Friction	3	Bread Machine	20
Animal Charcoal Machine	4	Brewing Apparatus	20 and 21
Antimony, 4; Anvil	4	Brick-making Machines	21
Aqueduct, 4; Arch	4	Bridges	21 to 28
Archimedean Screw	4	Buffer	28
Arming Press	4 and 5	Cables	28 and 29
Armour, 5; Arsenic	5	Cam, 29; Canal	29
Artesian Well	5	Candles	29 and 30
Artillery, 5 and 6; Assaying	6	Cement, 30; Chimney	30
Atomic Weights	6 and 7	Coal Cutting and Washing Machinery	31
Auger, 7; Axles	7	Coast Defence	31, 32
Balance, 7; Ballast	7	Compasses	32
Bank Note Machinery	7	Construction	32 and 33
Barn Machinery	7 and 8	Cooler, 34; Copper	34
Barker's Mill	8	Cork-cutting Machine	34
Barometer, 8; Barracks	8		

	Nos.		Nos.
Corrosion	34 and 35	Isomorphism, 68 ; Joints	68
Cotton Machinery	35	Keels and Coal Shipping	68 and 69
Damming	35 to 37	Kiln, 69 ; Knitting Machine	69
Details of Engines	37, 38	Kyanising	69
Displacement	38	Lamp, Safety	69, 70
Distilling Apparatus	38 and 39	Lead	70
Diving and Diving Bells	39	Lifts, Hoists	70, 71
Docks	39 and 40	Lights, Buoys, Beacons	71 and 72
Drainage	40 and 41	Limes, Mortars, and Cements	72
Drawbridge	41	Locks and Lock Gates	72, 73
Dredging Machine	41	Locomotive	73
Dynamometer	41 to 43	Machine Tools	73, 74
Electro-Metallurgy	43, 44	Manganese	74
Engines, Varieties	44, 45	Marine Engine	74 and 75
Engines, Agricultural	1 and 2	Materials of Construction	75 and 76
Engines, Marine	74, 75	Measuring and Folding	76
Engines, Screw	89, 90	Mechanical Movements	76, 77
Engines, Stationary	91, 92	Mercury, 77 ; Metallurgy	77
Escapement	45, 46	Meter	77, 78
Fan	46	Metric System	78
File-cutting Machine	46	Mills	78, 79
File-arms	46, 47	Molecule, 79 ; Oblique Arch	79
Flax Machinery	47, 48	Ores, 79, 80 ; Ovens	80
Float Water-wheels	48	Over-shot Water-wheel	80, 81
Forging	48	Paper Machinery	81
Founding and Casting	48 to 50	Permanent Way	81, 82
Friction, 50 ; Friction, Angle of	3	Piles and Pile-driving	82 and 83
Fuel, 50 ; Furnace	50, 51	Pipes	83, 84
Fuze, 51 ; Gas	51	Planimeter	84
Gearing	51, 52	Pumps	84 and 85
Gearing Belt	10, 11	Quarrying	85
Geodesy	52 and 53	Railway Engineering	85 and 86
Glass Machinery	53	Retaining Walls	86
Gold, 53, 54 ; Governor	54	Rivers, 86, 87 ; Rivetted Joint	87
Gravity, 54 ; Grindstone	54	Roads	87, 88
Gun-carriage, 54 ; Gun Metal	54	Roofs	88, 89
Gunnery	54 to 56	Rope-making Machinery	89
Gunpowder	56	Scaffolding	89
Gun Machinery	56, 57	Screw Engines	89, 90
Hand Tools	57, 58	Signals, 90 ; Silver	90, 91
Hanger, 58 ; Harbour	58	Stationary Engine	91, 92
Haulage, 58, 59 ; Hinging	59	Stave-making & Cask Machinery	92
Hydraulics and Hydraulic Machinery	59 to 63	Steel, 92 ; Sugar Mill	92, 93
		Surveying and Surveying Instruments	93, 94
Ice-making Machine	63	Telegraphy	94, 95
India-rubber	63	Testing, 95 ; Turbine	95
Indicator	63 and 64	Ventilation	95, 96, 97
Injector	64	Waterworks	96, 97
Iron	64 to 67	Wood-working Machinery	96, 97
Iron Ship Building	67	Zinc	96, 97
Irrigation	67 and 68		

In super-royal 8vo, 1168 pp., *with 2400 illustrations*, in 3 Divisions, cloth, price 13s. 6d. each; or 1 vol., cloth, 2l.; or half-morocco, 2l. 8s.

A SUPPLEMENT
TO
SPONS' DICTIONARY OF ENGINEERING.

EDITED BY ERNEST SPON, MEMB. SOC. ENGINEERS.

Abacus, Counters, Speed Indicators, and Slide Rule.
Agricultural Implements and Machinery.
Air Compressors.
Animal Charcoal Machinery.
Antimony.
Axles and Axle-boxes.
Barn Machinery.
Belts and Belting.
Blasting. Boilers.
Brakes.
Brick Machinery.
Bridges.
Cages for Mines.
Calculus, Differential and Integral.
Canals.
Carpentry.
Cast Iron.
Cement, Concrete, Limes, and Mortar.
Chimney Shafts.
Coal Cleansing and Washing.

Coal Mining.
Coal Cutting Machines.
Coke Ovens. Copper.
Docks. Drainage.
Dredging Machinery.
Dynamo - Electric and Magneto-Electric Machines.
Dynamometers.
Electrical Engineering, Telegraphy, Electric Lighting and its practical details, Telephones
Engines, Varieties of.
Explosives. Fans.
Founding, Moulding and the practical work of the Foundry.
Gas, Manufacture of.
Hammers, Steam and other Power.
Heat. Horse Power.
Hydraulics.
Hydro-geology.
Indicators. Iron.
Lifts, Hoists, and Elevators.

Lighthouses, Buoys, and Beacons.
Machine Tools.
Materials of Construction.
Meters.
Ores, Machinery and Processes employed to Dress.
Piers.
Pile Driving.
Pneumatic Transmission.
Pumps.
Pyrometers.
Road Locomotives.
Rock Drills.
Rolling Stock.
Sanitary Engineering.
Shafting.
Steel.
Steam Navvy.
Stone Machinery.
Tramways.
Well Sinking.

www.ingramcontent.com/pod-product-compliance
Lightning Source LLC
Chambersburg PA
CBHW032146230426
43672CB00011B/2468